Soybeans as a Food Source

Revised edition

Authors

W. J. Wolf

Research Leader, Meal Products
Oilseed Crops Laboratory
U.S. Department of Agriculture
Northern Regional Research Laboratory
Peoria, Illinois

J. C. Cowan

Adjunct Professor of Chemistry
Bradley University
(Formerly Chief, Oilseed Crops Laboratory
Northern Regional Research Laboratory)
Peoria, Illinois

CRC PRESS, INC.
Boca Raton, Florida 33431

Library of Congress Cataloging in Publication Data

Wolf, Walter James, 1927 —
   Soybeans as a food source.
   "Originally appeared as part of an article in CRC critical reviews in food technology."
   Bibliography: p.
   Includes index.
   1. Soy-bean as food. I. Cowan, John Charles, 1911 —   joint author. II. Title.
TX558.S7W65   1975   664'.8   75-9535
ISBN-0-8493-0112-2

This book represents information obtained from authentic and highly regarded sources. Reprinted material is quoted with permission, and sources are indicated. A wide variety of references are listed. Every reasonable effort has been made to give reliable data and information, but the author and the publisher cannot assume responsibility for the validity of all materials or for the consequences of their use.

All rights reserved. This book, or any parts thereof, may not be reproduced in any form without written consent from the publisher.

© 1975 by CRC Press, Inc.

Second Printing, 1977
Third Printing, 1979

International Standard Book Number 0-8493-0112-2
Former International Standard Book Number 0-87819-112-7

Direct all inquiries and correspondence to CRC Press, Inc., 2000 N.W. 24th Street, Boca Raton, Florida 33431

Library of Congress Card Number 75-9535
Printed in the United States

AUTHORS' INTRODUCTION

Recently the problems of world hunger have been discussed at length and an international awareness of the need for adequate nutrition has developed. Although malnutrition is found in many developing countries, it is not peculiar to them; affluent nations also have segments of their populations that suffer. Malnutrition may manifest itself as a lack of calories but often there is a deficiency of protein rather than of calories. Because animal proteins are expensive and in shortest supply where needed, plant proteins have received a great deal of attention. Lists of potential sources of unconventional food proteins rank soybeans at or near the top. Soybeans possess several advantages over other protein sources. They have a long history as a foodstuff in the Orient and supply protein, calories, and essential fatty acids. Soybeans are especially attractive as a protein source when compared to animal proteins on a cost basis. An important advantage of soybeans over other protein materials is their availability in large quantities. A variety of food-grade soy proteins are commercially available whereas other proteins often found in the potential category are still floundering with economic, technical, and legal problems. For example, fish protein concentrate has yet to succeed as a commercial venture in the U.S. and to become an established food ingredient (*Science*, 173, 410, 1971).

At present over 90% of the soybean oil utilized is consumed in foods. Indeed, soybean oil is the major food fat of the country. If large quantities of high linoleate-low saturate oils are needed to avoid heart disease in the United States (*Circulation*, 42, A55, 1970), soybeans represent the main available source. The essential linoleic acid serves as a precursor for prostaglandins, newly discovered (1964) hormones. These fatty acid hormones profoundly effect human cardiovascular and smooth muscle systems (*Ann. N.Y. Acad. Sci.*, 180, 218, 1971).

Although new developments will undoubtedly improve soy oil quality and extend use, soy proteins for foods represent a major relatively undeveloped outlet. Increased costs of animal proteins, improved quality of soy products, and more research commitments give impetus to this development. Further impetus is expected from an announcement that permits the use of textured vegetable (soy) proteins as meat extenders in the National School Lunch Program. Our review summarizes the conversion of soybeans to food. People newly concerned with soy proteins as a food but unfamiliar with the soy industry should find this monograph very helpful. Emphasis is given to the protein because of the present high level of interest.

W. J. Wolf
J. C. Cowan
Peoria, Ill.

THE AUTHORS

W. J. Wolf is Research Leader of Meal Products in the Oilseed Crops Laboratory, Northern Regional Research Laboratory, Peoria, Illinois.

Dr. Wolf received a B. S. degree in chemistry from the College of St. Thomas in 1950 and was trained as a biochemist at the University of Minnesota where he was awarded the Ph. D. degree in 1956. He then joined the staff at the Northern Regional Research Laboratory in Peoria as a research biochemist.

Dr. Wolf has authored more than 50 research reports and review articles on physical and chemical properties of soybean proteins and their uses in foods.

J. C. Cowan is former Chief of tne Oilseed Crops Laboratory, Northern Regional Research Laboratory, Peoria, Illinois. Presently he is adjunct professor of chemistry, Bradley University, Peoria, Illinois, and a consultant to industry.

Dr. Cowan earned an A. B. degree in chemistry at the University of Illinois in 1934 and was awarded the Ph. D. degree in organic chemistry at the same institution in 1938. He joined the Northern Regional Research Laboratory in Peoria as a research chemist in 1940 and retired in 1973.

Dr. Cowan's research interests have extended from polyamide and polyester resins now in commercial use to the many factors involved in the flavor stability of soybean oil. He has published over 250 papers and patents.

TABLE OF CONTENTS

Introduction 1

Seed Structure and Composition 2
    Structure 2
    Composition 3

Soybean Production 3
    Early History 3
    Areas of Production 5
    Production 6
    Importance of Varieties 6

Disposal of the Crop 7
    Grading Standards 7
    Disposition 7

Processing Soybeans into Oil and Meal 7
    Storage 8
    Preparation of Beans 11
    Extraction 11
    Desolventizing 12
    Degummed Oil and Lecithin Separation 12

Conversion to Edible Oil Products 12
    Alkali Refining 14
    Bleaching 14
    Hydrogenation 14
    Deodorization 15

Soybean Oil Products 18
    Salad and Cooking Oils 18
    Shortening and Margarine Oils 19
    Flavor Stability of Soybean Oil 22
    Soybean Lecithin — Products and Use 24

Food Uses of Soybean Proteins 25
    Physical and Chemical Properties 25
        Solubility as Function of pH 25
        Molecular Size 26
        Reactions of the 7S and 11S Globulins 27
        Solubility of Isolates 29
        Denaturation 30
        Amino Acid Composition 33
    Forms of Soy Proteins 34
        Whole Soybeans 34
        Processed Soybean Protein Products 34
    Selling Prices and Production Estimates 41

  Functional Properties 43
   Emulsification 43
   Fat Absorption 43
   Water Absorption 45
   Texture 46
   Dough Formation 48
   Adhesion, Cohesion, and Elasticity 50
   Film Formation 50
   Color Control 51
   Aeration 51
  Nutritional Properties 51
   Antinutritional Properties 51
   Protein Quality of Soybean Products 55
  Foods Containing Soy Proteins 59
   Oriental Foods 59
   Domestic Foods 60
  Problem Areas 65

Conclusions 69

Addendum 69

Introduction 69
  Origin of Soybeans 69
  Soybean Situation — Future 70
  Recent Sources of Information 71
  Soybean Organizations 72

Production 72
  Short-Term Situation 72
  Storage and Exports 74
  Soybean Varieties 75
  Yield Barrier 75
  Varieties and Antinutritional Factors 75
  Aflatoxin in Soybeans 76

Edible Oil Products 76
  Deodorization 76
  An Antioxidant for Soybean Oil 77
  Flavor Stability of Soybean Oil 77
  Oil from Field-Damaged Beans 78
  Flavor Components in Soybean Oil 79

Conversion to Edible Protein Products 79
  Production and Producers 79
  New Processes 80
   Full-Fat Products 80
   Defatted Flakes and Related Products 81
   Concentrates 81
   Isolates 82
   Textured Protein Products 82

Properties of Soy Proteins 83
   Functional Properties 83
      Solubility 83
      Water Absorption and Swelling 83
      Viscosity 84
      Emulsification 84
      Film Formation 84
      Texture 84
   Nutritional and Physiological Properties 85
      Trypsin Inhibitors 85
      Soybean Proteins in Blended Foods 85
      Nutritional Value of Textured Soybean Proteins 86
      Soybean Proteins in Infant Formulas 86
      Effect of Alkali Treatment on Soy Protein 86
   Flavor Studies on Soy Proteins 87
      Organoleptic Evaluation of Commercial Protein Products 87
      Origin of Flavor Compounds 87

Food Uses of Soybean Proteins 88
   Baked Goods 88
   Meat Products and Analogs 88
   Instant Breakfast Items 89
   Snack Foods 90
   Legal and Regulatory Aspects 90

References 90

Index 99

INTRODUCTION

Soybeans are a native crop of eastern Asia where they have served as an important part of the diet for centuries. The Japanese, for example, obtain 12 to 13% of their dietary protein from soybean products. For many of their traditional soy foods the Oriental people soak soybeans in water and then grind or cook them. Hot water extraction of ground beans yields soybean milk which is consumed as such or is treated with calcium salts to precipitate the protein plus oil in the form of bean curd or tofu. Fermentation of cooked soybeans yields products including soy sauce, miso, natto, and tempeh.

In the U.S., food uses of soybeans have followed a different pattern than in the Orient. Except for soy sauce, none of the traditional Oriental foods is consumed in significant amounts in this country. Soybeans are a relative newcomer to the American scene. They have only been grown in quantity since the late 1920's when soybean processing became an established industry; the two major products were oil and defatted meal. In the mid-1930's large portions of the oil began to be used for foods such as shortening, margarine, cooking oil, mayonnaise, and salad dressing. Because of its high protein content and good nutritional value, when properly processed, the meal was used primarily for animal feeds.

Soybeans have expanded in the last 30 years from a minor crop to a major cash crop. Indeed, in value to the farmer soybeans now rank second to corn and above wheat, potatoes, oats, cotton, and a variety of other crops better known to the consumer. Only within the last ten years, however, have very many edible products containing soybean derivatives been directly associated with their source. In shortenings, their presence was "hidden" by statements similar to the following: "A blend of hydrogenated vegetable oils," or in salad dressing, merely a "vegetable oil or a blend of vegetable oils." Today a long list of foods containing soybean-derived products can be prepared by careful reading of the labels in the supermarket. Yet most of these are even now not specifically identified as soybean. Products from corn, wheat, oats, and many other commodities are so labeled, for example, corn flakes, wheat germ, oatmeal, but not soybean. There are several reasons for this anonymity. Soybeans have a short history of use in the U. S. The flavor and texture of soybean products are comparatively strange to people outside the Orient. Although the Chinese and Japanese have converted soybeans into a variety of products, most of these foods have little physical or flavor identity with the original bean. Some people agree that green soybeans are a delicious dish when properly harvested and cooked, but their sale and the sale of mature beans for baking are extremely small. Soybean products have problems related to their flavor and flavor stability, to their function in foods, and to their physiological effects. Despite these problems soybean oils have become a major factor in our food industry. Soybeans now supply more than one half of the total visible fats and oils consumed in the U. S.

Meal consumption has paralleled the increase in oil usage and has assumed first place among protein supplements for livestock and poultry. Indeed, soybean meal has been a key factor in the enormous increase in poultry production as well as the comparatively low prices of chickens and turkeys. Similarly, meal has been an important factor in increased yields and lower costs in swine production. Meal has therefore served largely as an indirect source of edible protein.

Present outlets for soybean protein products such as flours, concentrates, and isolates are only a fraction of their potential outlet, but all increased many millions of pounds in the 1960's. Further substantial increases appear likely for the coming decade because the U.S. food processing industry is now undergoing major changes. Natural foods are being separated into their basic components-fats, carbohydrates, and proteins. These components serve as building blocks for new products, namely, engineered or fabricated foods. For example, milk has long been consumed directly but it is now also fractionated into sodium caseinate, lactalbumen, and lactose which serve as ingredients for the fabrication of new foods.

Because of their composition, soybeans are an excellent source of two components for fabricated foods—oil and protein. The extraction and refining of soybean oil is well established and soybeans now supply about 80% of the edible vegetable oils used in the U.S. In addition, the soybean industry has made significant progress in the last two decades in developing processes for converting soybeans into a variety of products with protein contents ranging from about 40% for full-fat flours to 95% or more for protein isolates. Soybeans also have the advantage of being available in large quantities. They are grown on over 40 million acres in the eastern half of the U.S. from Minnesota in the north to Louisiana and Florida in the south and from central Kansas to Delaware. Since 1967, the crop has been over a billion bushels annually.

Food-grade isolates appeared on the market in 1959 but uses developed slowly despite the obvious nutritional advantages of fortifying foods with these proteins. It was generally accepted by the manufacturers of soy proteins and food companies alike that "you can't sell nutrition." As a result, markets for the isolates and protein concentrates (which also entered the markets in 1959) were developed primarily on the basis of their functional properties such as emulsification, thickening, and fat-binding. This situation, however, has changed considerably in the last two years. The White House Conference on Food and Nutrition in 1969 and U.S. Senate subcommittee hearings in July-August 1970 on breakfast cereals have pinpointed foods which the industry could improve nutritionally by fortification with proteins and other nutrients. There is now considerable activity in the development of ingredients and foods based on soy proteins by companies with varied backgrounds cereals, meats, dairy products, soft drinks, pharmaceuticals, and chemicals. Two companies recently entered this field by acquiring smaller concerns specializing in soy protein items. Development work and test marketing are actively underway. For example, companies supplying ingredients for snack items now have available expandable starches which provide a matrix for addition of proteins such as soy isolates and concentrates.

Our review summarizes the current status of food uses of soybean oil and protein in the U.S. Data on soybean production, disposal of the crop, seed composition, and structure are included as background for an overall view. Problem areas and future trends in usage of soybean oil and protein are discussed.

SEED STRUCTURE AND COMPOSITION

Structure

Soybeans are typical legume seeds which differ in size, shape, and color depending on the variety.

They range from small round beans to large, oblong, flattened seeds of yellow, brown, green, or black or a combination of these colors. The common field varieties grown in the U.S., however, are nearly spherical and yellow. The Lincoln soybean, formerly one of the major varieties grown in Illinois, Iowa, and Indiana, is shown in Figure 1. A cross section of a soybean (Figure 2) shows the two major structural parts — the hull and cotyledon. Two minor structures — the hypocotyl and plumule are not shown. Proceeding inward, the hull is made up of an outer layer of palisade cells, a layer of hourglass cells, smaller compressed parenchyma cells, aleurone cells, and finally compressed layers of endosperm cells. The surface of the cotyledon is covered with an epidermis, and the interior is filled with numerous elongated palisade-like cells filled with protein and oil. A more detailed view of the cells in the interior of the cotyledon is shown in Figure 3. The bulk of the proteins are stored in the protein bodies which may vary from 2 to 20 μ in diameter.[3] The oil is located in the smaller structures called spherosomes which are interspersed between the protein bodies and are 0.2 to 0.5 μ in diameter. The protein bodies survive grinding of the seed and have been isolated from defatted soybean flour.[3,4] Isolated protein bodies may contain as high as 98% protein, and a minimum of 60 to 70% of total protein in defatted flour appears to be "prepackaged" in these subcellular structures.[3] Theoretically, it appears that methods of milling and separation should be possible whereby the protein bodies would be separated in a reasonably pure state on a commercial scale. Attempts to separate a protein-rich fraction from defatted soybean flour by air classification were unsuccessful;[5] perhaps a different approach to the preparation of defatted flour is needed.

Composition

Commercial soybeans constitute approximately 8% hull, 90% cotyledon, and 2% hypocotyl and plumule. Proximate compositions for whole beans and three fractions are given in Table 1. The constituents of major interest—oil and protein—make up about 60% of the bean, but about one third consists of carbohydrates including polysaccharides, stachyose (3.8%), raffinose (1.1%), and sucrose (5.0%).[6] Phosphatides, sterols, ash, and other minor constituents are also present.[7] Oil and protein contents depend on variety, soil fertility, and weather conditions. Composition of the protein and oil are discussed later.

SOYBEAN PRODUCTION

Early History

The soybean was grown in the U.S. to only a limited extent prior to 1920. Initially, most of the acreage was used for forage and pasture; only limited amounts were grown and harvested for

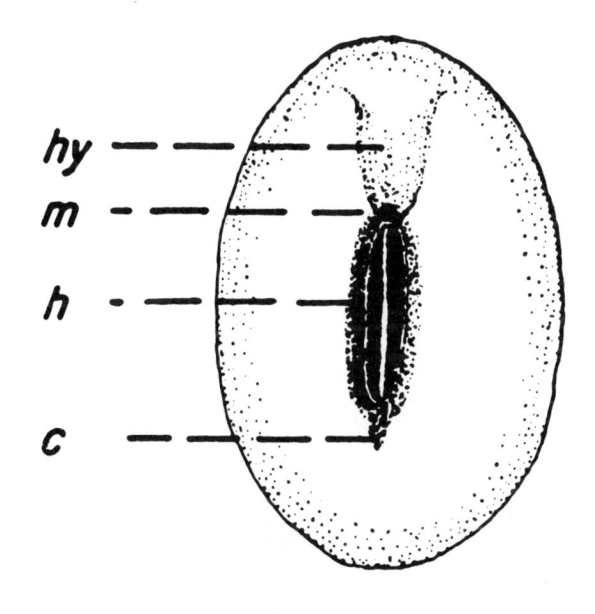

FIGURE 1. Drawing of Lincoln variety soybean. Structural features: h = hilum or seed scar; c = chalaza; m = micropyle; and hy = outline of hypocotyl under the seed coat. (From Williams.[1])

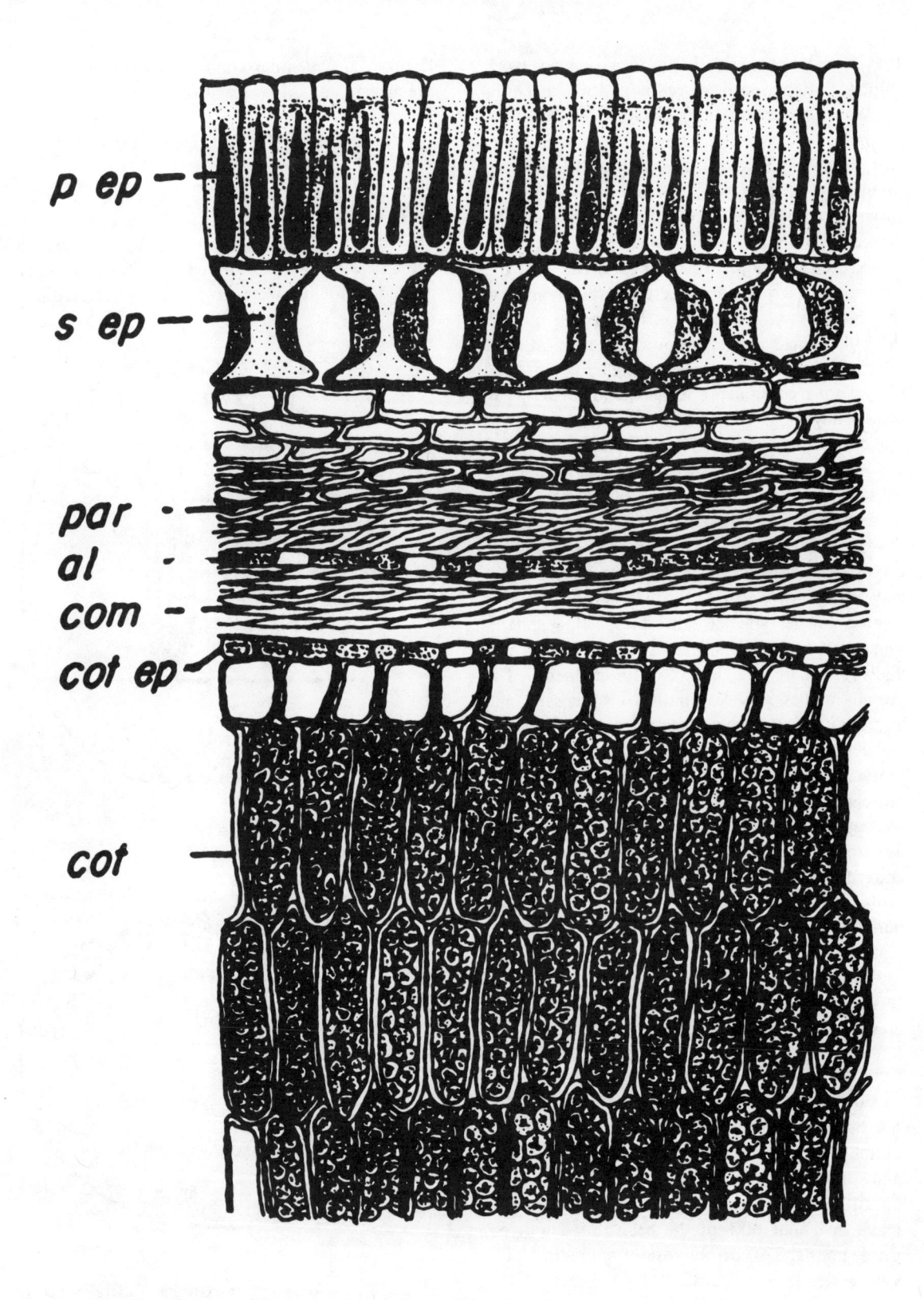

FIGURE 2. Cross section of soybean hull and part of the cotyledon. Structures in the hull: *p ep* = palidade cells; *s ep* = hourglass cells; *par* = parenchyma; *al* = aleurone cells; and *com* = compressed cells of the endosperm. In the cotyledon *cot ep* = the epidermis and *cot* = the aleurone cells. (From Williams.[1])

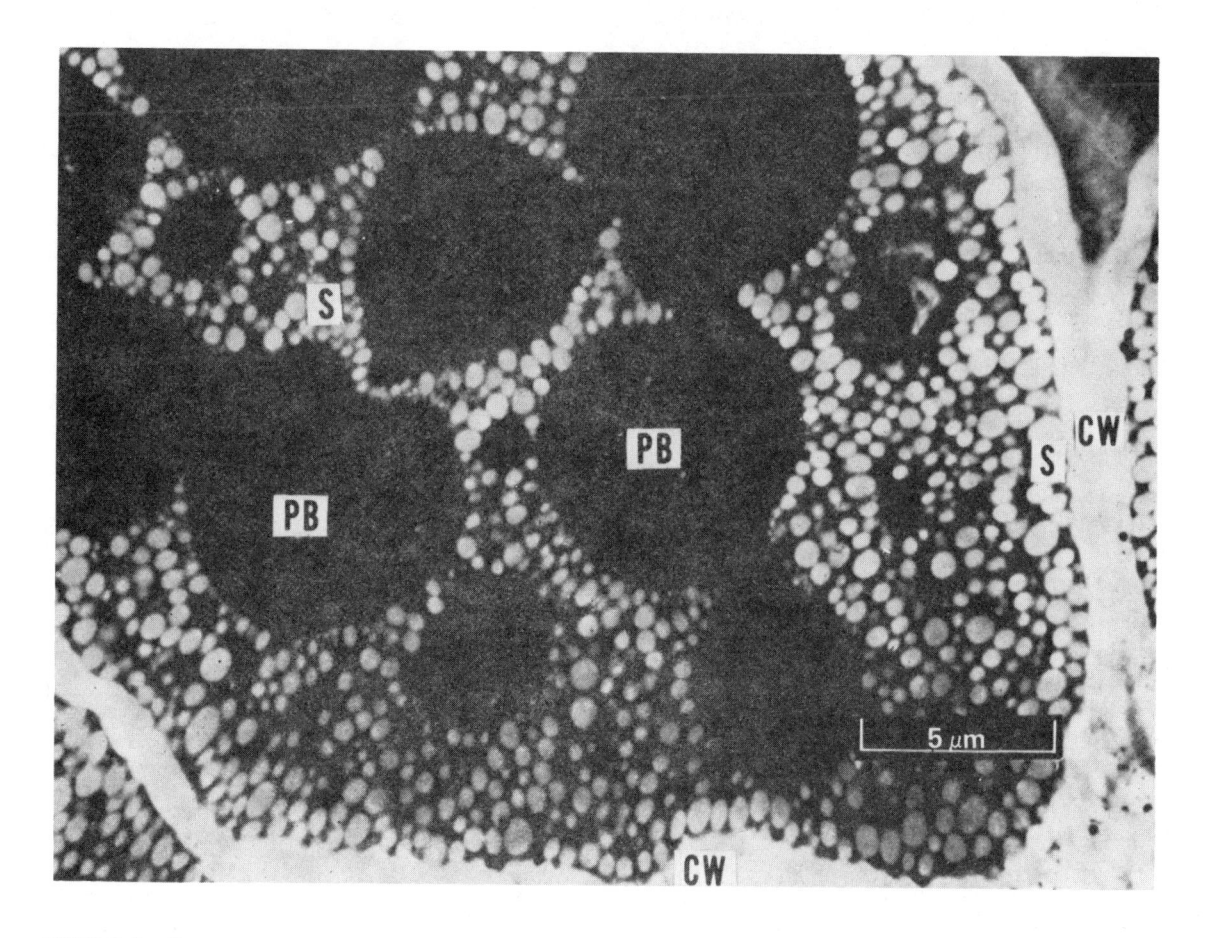

FIGURE 3. Transmission electron photomicrograph of section of mature soybean cotyledon. PB = protein bodies; S = spherosomes; and CW = cell wall. (From Saio, K. and Watanabe, T., *J. Food Sci. Technol.* Japan, 15, 290, 1968. With permission.)

TABLE 1

Proximate Composition of Soybeans and Seed Parts*

Fraction	Protein (N × 6.25), %	Fat, %	Carbohydrate, %	Ash, %
Whole bean	40	21	34	4.9
Cotyledon	43	23	29	5.0
Hull	8.8	1	86	4.3
Hypocotyl	41	11	43	4.4

*Moisture-free basis.
(From Kawamura.

seed. Processing of domestic soybeans for oil began in 1915 and seed production was thereby stimulated so that by 1920 a three-million-bushel crop was produced. At that time North Carolina, Virginia, Alabama, Missouri, and Kentucky were the leading producers of soybeans. In 1922 the A. E. Staley Manufacturing Co., a major corn-processing concern in Decatur, Illinois, began to crush soybeans. During these years soybean production expanded in the North Central states and by 1924 Illinois became the major producer. Illinois has been the leader in acreage and production ever since. Decatur subsequently became the major soybean processing center in the U.S. and is the base point in pricing soybeans.

Problems faced by the industry in the beginning were: (a) to obtain enough soybeans from farmers for economical processing; (b) to develop efficient processing methods; and (c) to find users of soybean oil and meal. These problems were largely met by contracts between processors, growers, and a feed company to purchase all beans produced on a given acreage, and by the mid-1930's the industry was firmly established.

Areas of Production

The Corn Belt states account for the majority of the production of soybeans in the U.S. Table 2 lists the ten major producing states. Illinois and

TABLE 2

Soybean Production in the Ten Leading States for 1970

State	Acreage (million acres)	Yield (bushels/acre)	Production (million bushels)
Illinois	6.66	31.0	206.5
Iowa	5.49	33.0	181.3
Indiana	3.25	32.0	103.8
Arkansas	4.23	23.5	99.4
Missouri	3.50	26.0	90.9
Minnesota	3.10	26.5	82.3
Ohio	2.44	29.5	71.9
Mississippi	2.34	25.0	58.4
Louisiana	1.67	23.0	38.5
Tennessee	1.21	23.0	27.7
Total U.S.	41.62	27.3	1,134.2

(From *Fats and Oils Situation*.[8])

Iowa are the two leading states because of the large acreages harvested and high yields. Arkansas and Missouri, fourth and fifth in production, each harvested more acres than Indiana but were out-produced by Indiana because of its high yield per acre. Production in the Southern states has been on the increase because of declining cotton acreage over the last 20 years. In Arkansas, for example, soybean acreage has increased sevenfold and production eightfold since 1950.

Production

Soybeans have shown a phenomenal growth as a crop in the U.S. From a three-million-bushel crop in 1920, production has increased 378-fold in 50 years. Table 3 shows production by five-year intervals for 1925-1970. Two factors are responsible for the present size of the crop--a 24-fold increase in acreage and a 2.3-fold gain in yield per acre. Further improvements in yield can be expected as more is learned about the physiological and agronomic characteristics of soybeans and as present knowledge is applied by the growers. The 1.13-billion-bushel U.S. crop for 1970 is about 75% of the 1.5-billion-bushel world crop. Food uses of soybeans as a protein source in this country are estimated to be only about 10 million bushels annually or less than 1% of the crop. Thus, there is no problem of supply for the food market.

Importance of Varieties

Soybeans show a marked response to photoperiod, and each variety has a characteristic length of day at which blooming and seed development begin. Since day length varies with latitude, a given variety is generally best adapted to narrow bands (100 to 150 miles wide) running east and west. Consequently, no single variety dominates the market. Agronomists have developed a number of varieties suited to the different regions in which soybeans are grown. Soybean varieties are divided into ten maturity groups designated 00 through VIII. Group 00 varieties are adapted for the northern latitudes--southern Canada, northern Minnesota, and North Dakota--while Group VIII varieties are designed for the southernmost region--the Gulf Coast. Varieties recommended in 1970 for the major growing areas included: Portage, Merit, Traverse, and Chippewa 64 for the

TABLE 3

United States Soybean Acreage and Production 1925–1970

Year	Acres harvested (million acres)	Yield (bushels/acre)	Production (million bushels)
1925	0.4	11.7	5
1930	1.1	13.0	14
1935	2.9	16.8	49
1940	4.8	16.2	78
1945	10.7	18.0	193
1950	13.8	21.7	299
1955	18.6	20.1	374
1960	23.7	23.5	555
1965	34.5	24.5	846
1970	41.6	27.3	1,134

(From *Fats and Oils Situation*[8] and *Soybean Digest*.[9])

northern regions; Amsoy, Wayne, Harosoy 63, Beeson, Clark 63, and Calland for the Midwest: and Hill, Lee, Hood, and Dare for the South. Further details are found elsewhere.[9]

Plant breeding for desired characteristics such as yield, disease resistance, and composition is a continuous program; hence, the major varieties grown keep changing. Several high-protein varieties have been introduced in recent years but as far as we are aware they are not being grown specifically for food uses. Field-run beans of No. 1 or No. 2 yellow grade (see Table 4 for definition of grades) are used in the preparation of soy protein products such as isolates. No varieties are segregated for this purpose since there is still no clearly defined correlation between varietal characteristics and properties of the resulting protein products. Segregation of varieties, however, is carried out on a limited scale for export to Japan were certain varieties are preferred for some of their traditional soybean foods.

DISPOSAL OF THE CROP

Grading Standards

Soybeans are classified as cereal grains; hence, trading is regulated by the U.S. Grain Standards Act. The Act is administered by the USDA in accordance with the *Handbook of Official Grain Standards of the United States*. Classification of soybeans is according to color, and yellow soybeans constitute the major commercial class. Grades are based on test weight, moisture content, and percentages of splits, damaged kernels, and foreign material. Table 4 lists requirements for numerical and sample grades of soybeans.

Disposition

The major portion of the crop is processed into oil and meal while the remainder is largely exported. For 1969, crop disposal was as follows:

Use	Million bu	% of supply
Processing	738	51
Export	429	30
Seed	49	3
Carryover	229	16

The soybean processing industry has expanded steadily over the years with increases in production. Annual capacity late in 1970 was estimated to be 825 million bushels and was predicted to

TABLE 4

Grading Requirements for Soybeans

Requirements	No. 1	No. 2	No. 3	No. 4
Min test wt, lb/bu[†]	56	54	52	49
Moisture, %	13	14	16	18
Max limits, %				
Splits	10	20	30	40
Damaged kernels	2	3	5	8
Foreign material	1	2	3	5
Colored beans[‡]	1	2	5	10

*Soybeans not meeting requirements of grades 1 to 4, inclusive, or which are musty, sour, or heating; or which have an objectionable odor are *Sample grade*.
[†]Legal weight for sale purposes is 60 lb/bu.
[‡]Not yellow or green.
(From Official Grain Standards of the U.S.[10])

increase to about 900 million bushels by the spring of 1971. The industry has grown from a large number of small, inefficient mills to a small number of large, efficient units. At present about 130 mills process an average of 6.0 million bushels a year as compared to 3.2 million bushels in 1960. The largest mills are found in the North Central Region—mills in Illinois processed an average of 13 million bushels in 1969.[8]

Exports of soybeans have increased rapidly during the last 15 years. Greater demands for oil and meal and favorable processing margins and capacities in Western Europe and Japan suggest that exports will continue as long as soybeans are competitive in price. In the 1969 crop year soybeans and soybean products contributed substantially to the U.S. balance of trade with soybeans alone amounting to $1.14 billion and exported oil and meal adding another $0.4 billion.

PROCESSING SOYBEANS INTO OIL AND MEAL

Processing—a necessity for the most effective use of soybeans—removes the oil which is used by the edible fat industry and converts the defatted meal into feeds and food products. Soybean meal contains factors that must be inactivated by moist heat before optimum growth rates are obtained with young animals when the meal is used as a feed. For food uses the processing may consist of merely heating and grinding the defatted material as in the preparation of flours and grits or of further fractionation to increase protein content as

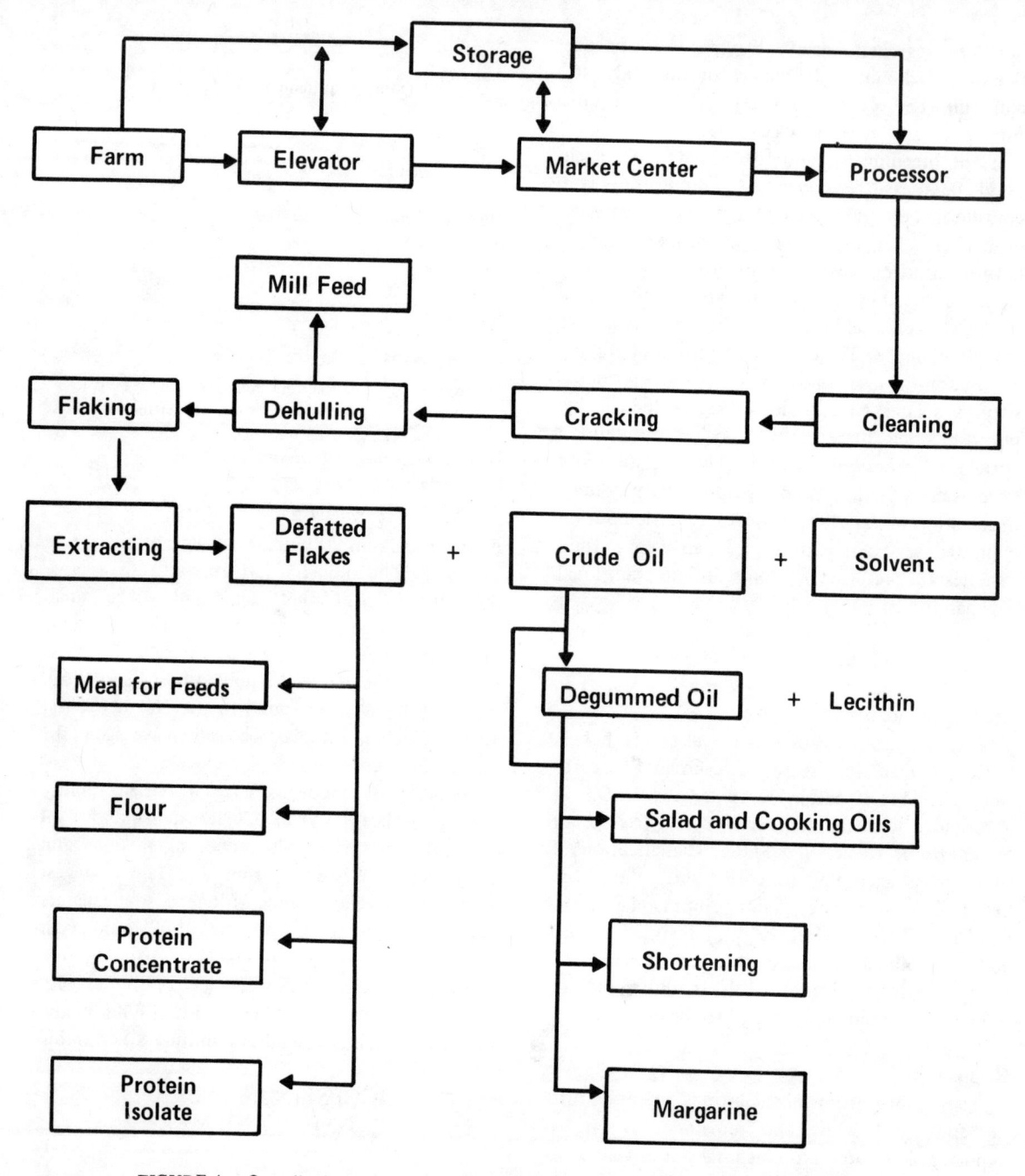

FIGURE 4. Overall scheme for processing soybeans into oil, flakes, and derived products.

in the production of concentrates and isolates. Figure 4 outlines the overall scheme for converting soybeans into oil, meal, and related products by solvent extraction. The processing industry has evolved from hydraulic presses to mechanical screw presses or expellers to solvent extraction. The present industry is close to 100% solvent extraction.

Storage

Soybeans are unique when compared to other major U.S. farm grains; about 95% of the crop is marketed for processing or exporting and processing overseas. The orderly flow of soybeans from the farm to the marketplace thus begins with storage. The U.S. has an annual processing capacity of about 900 million bushels, but production of oil and meal must be spaced over 10 to 12 months. As a result elevators, many farmers or marketing cooperatives, and grain terminals have large storage facilities. All large processors have substantial storage capacity to maintain supplies for processing. Many mills can store as much as five to seven million bushels. With some plants milling and extracting as much as 1200 tons a day, they will consume 40,000 bushels a day. If all of the storage available were devoted to soybeans, these plants would have several months' available. On-the-farm storage or storage relatively close to the actual production area is substantial because prices for soybeans tend to be lower at the beginning of the marketing year and higher 6 to 11 months later.

Concrete silos with diameters of 20 to 40 ft and heights above 150 ft serve as commercial storage. These silos are usually arranged in double to triple rows, and considerable additional storage is gained from interstitial spaces. The beans are cleaned and dried if necessary before they are stored (Figure 5). Soybeans are usually dry enough that storage is not a major concern in most of the processing industry, but proper moisture content is needed for successful storage. With an initial 12% moisture, beans will usually store without change in grade for 2 years. At 13 to 15% moisture, beans can be stored successfully for several months during cool weather. High-moisture beans (14% or more) are usually dried before storage.

Migration of moisture from warm areas to cooler surfaces in the bins can occur and cause localized increases in moisture that, in turn, cause localized heating. Most commercial storage bins have temperature monitoring systems. When any sign of local overheating occurs, it is common commercial practice to move these soybeans into another bin or to processing. This movement mixes beans from the higher moisture areas with beans from the lower moisture areas and controls heating. The increase in temperature can be great enough to char the beans so much that they are black and burned. Indeed, the beans can catch on fire and the whole bin or more lost.

High moisture content promotes mold or fungus growth that can impart a moldy odor to reduce the beans to sample grade. Table 5 shows the effect of different levels of moisture on short-term storage of soybeans at elevated temperatures. Since the discovery of aflatoxins in peanut meal in the early 1960's, concern about their presence in soybeans and soybean meals was raised. Examination of 866 soybean samples during two crop years gave only two samples containing aflatoxin.[12] Both were sample grade and showed obvious damage. Attempts to produce aflatoxin on soybeans with aflatoxin-producing strains of *Aspergillus flavus* failed to produce much toxin.[13] Thus, it is improbable that foods would become contaminated with aflatoxin from soybeans for two reasons: 1) sample grades are not likely to be used for food processing, and 2) moldy soybeans do not produce much toxin.

In long-term storage of soybeans, nutritive value can be lowered more than in wheat or corn. Only a very small part of the wheat or corn kernel is germ, whereas the whole bean is mostly living germ. Soybean meal products under proper storage conditions suffer little loss in nutritive value when stored as long as three years.[14]

Since soybeans are easy to store and have a high bulk density (45 lb per cu ft), they may be shipped far from the growing areas and even

TABLE 5

Effect of Moisture Content on Soybeans Stored at 37.8°C for 11 Days

Initial Moisture %	Respiratory rate, mg CO_2/100 g/24 hr	Acid value of oil*	Germination %
11.8	0.5	1.0	85
13.0	0.9	1.1	74
15.1	0.9	1.5	43
16.0	5.4	1.8	15
18.3	73.3	6.9	0

*Mg KOH to neutralize 1 g of oil.

(From Milner and Geddes.[11])

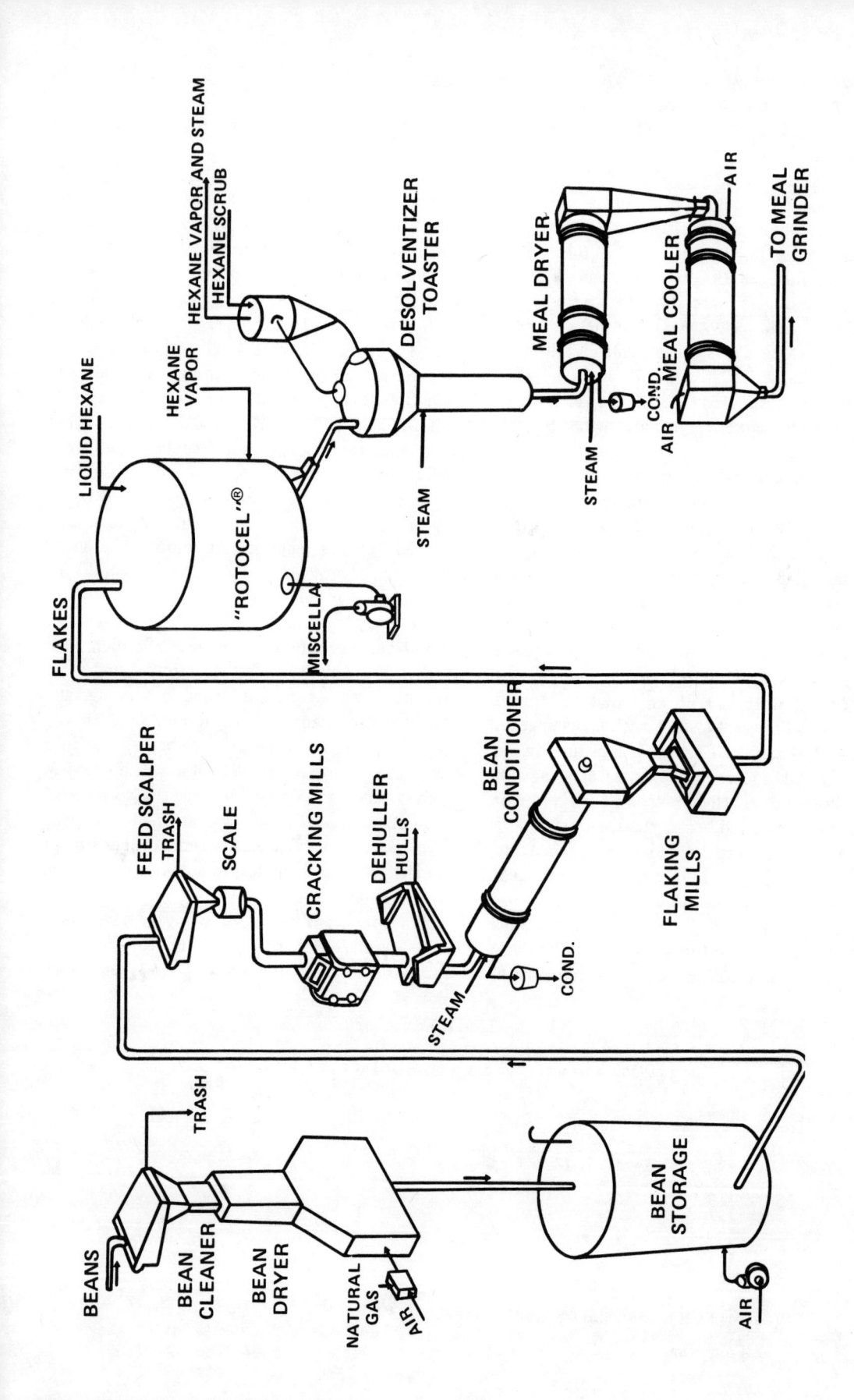

FIGURE 5. Flow chart for a soybean solvent extraction plant. (From Blaw-Knox Chemical Plants, Inc. With permission.)

exported to Europe and the Orient. For marketing reasons, much of the soybean crop is processed relatively close to the production areas or in close proximity to regions where livestock and poultry feeding are carried out.

Preparation of Beans

All soybeans used to prepare defatted products are treated in a similar manner up to the flaking operation.[15] Recent work on inactivation of enzymes for soy beverages[16] and other full-fat products may result in some changes through the use of moist heat on the beans. Generally, the beans move from storage into a feeding hopper that controls their flow. They are cleaned by passing through a magnetic separator and a screening operation. The separator removes iron, steel, and other magnetically susceptible objects. Shaking operations with different sizes of screens separate the larger and smaller sized seeds from the soybeans.

Next, the cleaned beans flow to the cracking rolls (Figure 5). These spiral-cut corrugated cylinders usually have 6 to 10 cuts per inch on rolls that are 14 to 16 in. in diameter and 48 in. wide. Cracking into six to eight pieces and loosening of the hulls are effected by passing the beans through two or three pairs of rolls. One roll in each pair travels at a different speed than the other to effect cracking. For food and high-protein feeds for poultry, dehulling is usually practiced after cracking to reduce fiber content. Dehulling is also carried out to improve efficiency of the extraction plant since the hulls contain very little oil. Hulls are removed by aspiration of the cracked beans. The hulls may be toasted separately to give products known as mill run and mill feed or may be returned to the processing stream after the extractor to make 44% protein meal; dehulled, defatted meal contains a minimum of 49% protein (12% moisture basis).

The cracked beans are conditioned to 10 to 11% moisture at 63 to 74°C and then flaked by passing through smooth rolls (20 to 32 in. in diam, 42 to 48 in. long) traveling at different speeds. Strong springs are used to exert proper pressure on the rolls to ensure relatively thin flakes. The clearance is usually adjustable so that flakes of uniform thickness may be obtained. Flakes should be about 0.01 to 0.015 in. thick. In this range a threefold increase in thickness slows extraction rate ~80 times. Flaking ruptures the cells in the soybean and reduces the distance that oil and solvent must diffuse, thereby facilitating extraction with organic solvent.

Extraction

The choice of solvents in the extraction of soybeans for oil and meal is based primarily on the ease with which the oil is removed from the flakes. In the United States, the petroleum hydrocarbon mixture called hexane (bp 66 to 69°C) is widely used. A similar mixture called heptane (bp 89 to 98°C) is sometimes used in climates where outside temperatures are high in order to reduce explosion hazards. Cyclic hydrocarbons find favor in Europe where a mixture (bp 71 to 85°C) composed primarily of cyclohexane is used. Ethanol[17] and trichloroethylene[18] have been used commercially. Alcohol improves flavor and lowers functionality of soybean protein products.[19] Trichloroethylene is a much safer solvent to handle to avoid explosions and fires, but it reacts with the soybean flakes to produce products toxic to cattle and sheep.[20] This chlorohydrocarbon reacts with cysteine to give S-dichlorovinylcysteine that causes aplastic anemia and disease symptoms in cattle that are identical to the effect of the trichloroethylene-extracted meal when fed to cattle. Use of this solvent with food or feed products containing sulfhydryl groups should be avoided.

The petroleum hydrocarbons used in the United States for the extraction of soybean flakes are processed to remove any toxic materials from them that might be left in the flakes.[21] Although some hydrocarbon solvents used or found in research laboratories may contain measurable amounts of carcinogenic hydrocarbons,[22] the extraction grade of hexane has no measurable amount by the available methods.

The soybean flakes flow by conveyor to the extractor (Figure 5). Two basic types of extractors were originally used in the United States: Hansa-Mühle or Bollmann extractor, which is a series of baskets mounted on a vertical endless chain, and a U-tube or vertical immersion extractor known as the Hildebrandt. These extractors were introduced in the 1930's and some of the more recent models are still running. Currently, extractors employ a variety of different ways to contact the flakes, including a presoaking period to remove some of the oil and stationary baskets with the pumping of solvent and miscella in a progressive stepwise countercurrent flow of flakes and solvent.[15]

The miscella is filtered to remove fines and the solvent is stripped from the crude oil by passage through preheaters, using film evaporators and stripping columns. The latter are usually packed, and they are steamed countercurrently and maintained under diminished pressure to remove the last of the hexane.

Desolventizing

For feed use the hexane-laden flakes are passed through a desolventizer-toaster which recovers the hexane and simultaneously toasts the flakes to obtain optimum nutritive value (Figure 5).[23] The desolventizer-toaster is a vertical series of steam-jacketed compartments with a revolving center shaft equipped with blades. The blades move the flakes around until they reach an opening in the floor of the compartment and drop into the next lower compartment. Steam is introduced to vaporize the hexane and condense on the flakes to increase their moisture content. As the flakes descend through the unit, temperature is gradually increased to 110°C to lower the moisture to 13 to 15% at the end of the cooking cycle. After drying and cooling, the flakes are ground into meal. For food uses the hexane-containing flakes are processed differently as described later (Forms of Soy Proteins - Soy Flours and Grits).

The material balance for the extraction operation is shown in Figure 6.

Degummed Oil and Lecithin Separation

Crude oil from the stripping columns may be dried and sent directly to the edible oil refinery. It may also be treated to remove phospholipids to furnish soybean lecithin to the food and pharmaceutical industry. About 1% moisture is added to the warm oil in a suitable mixing chamber and the hydrated phosphatides are continually removed in pressure or other centrifuges. The resulting product contains about 25% moisture with the remainder oil and phospholipid. This material is dried under reduced pressure at 65 to 70°C to lower moisture to less than 0.5%. This soybean lecithin contains about 30% oil and 70% phospholipids. The degummed oil contains a maximum of 0.02% phosphorus, 1.5% unsaponifiables, 0.75% free fatty acids, and 0.3% of moisture, volatile, and insoluble impurities by definition.

The phospholipids may also be separated from crude oils by use of acetic anhydride. About 0.1 to 1% of anhydride is mixed thoroughly with the crude oil and warmed to 60°C. About 1.5% water is added with mixing and the aqueous phospholipid layer is separated by centrifugation. The oil is washed with water and centrifuged again to give a break-free oil suitable for bleaching and deodorizing.[24] Degummed oil from treatment of crude oil with water alone must be alkali-refined.

CONVERSION TO EDIBLE OIL PRODUCTS

The flow scheme (Figure 7) for edible oil products furnishes an overall view of the manufacture of shortening, margarine, and cooking oils. With soybean oil, a bland flavor and clear and brilliant appearance when liquid are usually required. Generally, degummed soybean oil is alkali-refined, bleached, and deodorized for use in salad dressings; the bleached oil is hydrogenated, winterized, and deodorized for use as a salad and cooking oil and liquid shortening. For some salad oil outlets, the bleaching step is omitted. The bleached oil may be mixed with other bleached oils for hydrogenation to shortening stock that may be further blended with other shortening stocks. Blended oils are hydrogenated to give margarine stocks that are usually blended again to give the proper plasticity in the final product. With recent "tub" margarines, a hydrogenated oil is

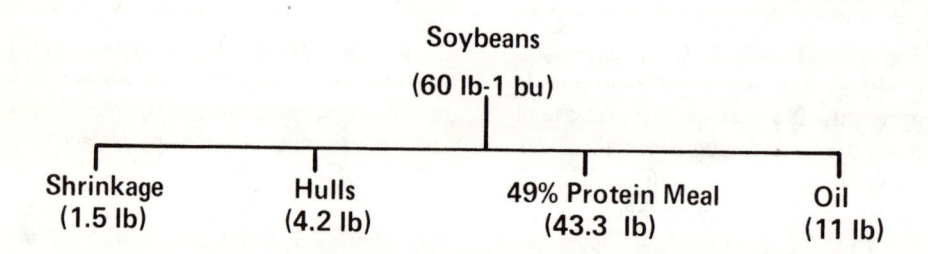

```
                    Soybeans
                  (60 lb-1 bu)
                        |
    ┌───────────┬───────┴────────┬──────────┐
 Shrinkage    Hulls      49% Protein Meal   Oil
 (1.5 lb)   (4.2 lb)        (43.3 lb)     (11 lb)
```

FIGURE 6. Material balance for solvent extraction of soybeans.

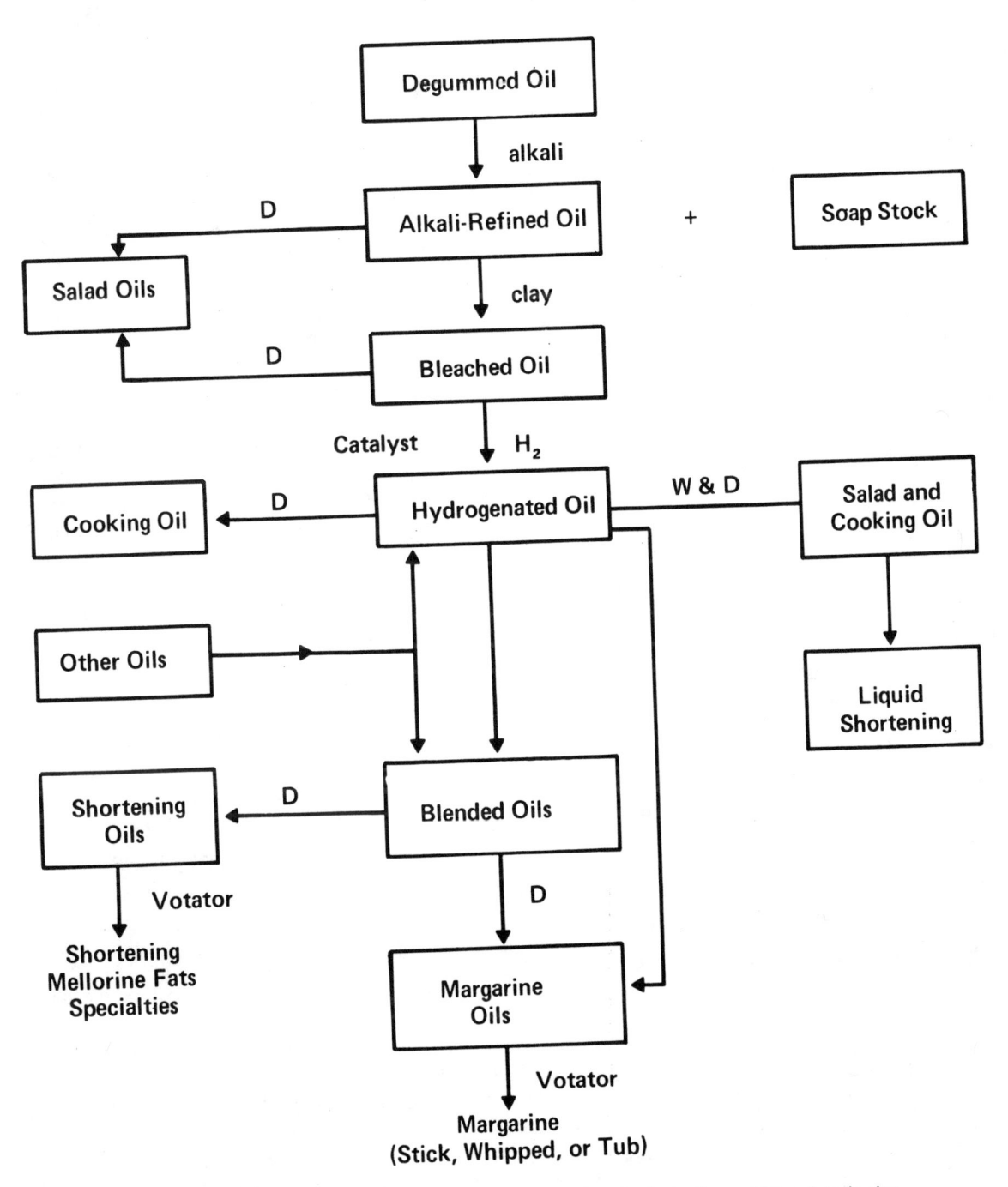

FIGURE 7. Flow scheme for edible oil products, where D = deodorization and W = winterization.

blended with a liquid oil. Thus, the manufacturer of edible oil products draws on a number of processes to improve the physical and flavor characteristics of his product and makes considerable use of blending to achieve his goals. Blending usually occurs just prior to deodorization but may occur afterwards.

Alkali Refining

Neutral oil with a free fatty acid content of less than 0.05% as oleic acid is required before the operations of bleaching, hydrogenating, and deodorizing are effected to give consumer products. Alkali refining is usually conducted on a continuous basis by one of the following procedures: caustic soda (sodium hydroxide), soda ash (sodium carbonate), or a combination of these two. Refining may also be done with acetic anhydride on the crude oil and a combination of phosphoric-citric acid-lye. Caustic soda and degummed or crude oil are continuously metered into a motor-driven mixer. The oil is warmed to 60 to 70°C by passage through a steam heater. A battery of centrifuges separates the soapstock from the oil. Final traces of alkali are removed by washing with 10% water at 70 to 80°C. A second washing is often used with some citric acid in the wash water. Mixing with alkali and centrifuging requires about three minutes and water washing requires an additional ten minutes. The washed oil may be dried by spraying into a vacuum of about 4 mm to reduce moisture from about 0.5 to 0.05%. For further details, see Norris[25] and Sullivan.[26]

Bleaching

Most soybean oils are improved in flavor stability if they are bleached prior to deodorization.[27] Bleaching is also desirable for stocks that go to the hydrogenator. Natural bleaching earths, such as fuller's earth (which is a hydrated aluminum silicate), acid-activated clays composed mainly of bentonite or montmorillonite, and activated carbon, are used to bleach soybean oil. Since soybean oil is usually easy to bleach, natural earth or activated clays are generally used. If clay fails to achieve the desired color, a mixture of one part of activated carbon in 10 to 20 parts of clay is often utilized as an adsorbent for hard-to-bleach oils.

Continuous vacuum bleaching is commonly used in the industry.[25] Deaerated and dried oil is continuously mixed with the clay. The resulting slurry is sprayed into an evacuated chamber to remove air and unbonded moisture in the clay. The slurry is heated to 105 to 110°C for bleaching and again sprayed into an evacuated chamber to remove any bonded moisture released by heating the clay. The oil and clay are separated in a closed filter press. Countercurrent bleaching can be effected by passing dried, deaerated, and heated oil through the filter press containing the clay from a previous bleaching operation. The oil then goes to the process or treatment vessel for the second bleaching operation where fresh clay is continuously mixed with the oil at bleaching temperature. The clay-oil mixture flows to a second-stage filter for separation. With a third filter available, one can always be free for cleaning and readying for service to make the process continuous. The amount and kind of adsorbent will depend on the characteristics of the incoming oil and the degree of bleaching desired. With activated clays, less than 1% adsorbent is usually used unless the oil is green or difficult to bleach. Green oils occur when immature or frost-damaged beans are processed. Generally, carbon black mixed in with the clay will remove the green color. Results with the three types[28] of bleaching procedures—batch open kettle, batch vacuum, and countercurrent vacuum—are given in Table 6. Bleached oils flow directly to either the hydrogenator or to the deodorizer or to both in sequence for conversion into edible products.

Hydrogenation

Much of the soybean oil used for food purposes is hydrogenated to form margarine base oils, shortening, and cooking oils as well as the all-purpose salad and cooking oil.[29] Nickel is the preferred catalyst. It can be prepared from nickel formate by decomposition at 400 to 450°F in two to four parts of oil, or by precipitating nickel hydroxide on kieselguhr from nickel nitrate solution with alkali, or by electrolytic precipitation. The catalyst is activated by heating to 900 to 1000°F under hydrogen. Platinum or palladium catalysts may also be used but are usually more expensive.

The addition of hydrogen to the unsaturated bonds with any of these three catalysts appears to be a complex reaction. It is thought to proceed by the Horiuti-Polanyi mechanism.[30] A double bond adsorbed on the surface of the catalyst adds an atom of activated hydrogen. A second atom of hydrogen many be added to saturate the double

TABLE 6

Bleaching of Soybean Oil with Acid-Activated Clay

Procedure	% Clay	Lovibond color red Initial	Final	% Free fatty acids* Initial	Final	AOM[†] stability (hr)
Batch open kettle	1.0	11.5	2.8	0.12	0.15	9
Batch vacuum	1.0	11.5	2.1	0.12	0.19	11
Countercurrent	0.8	11.5	1.9	0.12	0.14	14

*Free fatty acids as % oleic.
[†]Active oxygen method.

(From Singleton and McMichael.[28])

bond, or a molecule of hydrogen is abstracted by the catalyst to give the same or new double bond. Thus, both position and *cis-trans* isomerizations have occurred when the fatty molecule is released from the catalyst. Palladium effects more isomerization than nickel or platinum catalysts. With oleic ester and deuterium, the unsaturated ester that remained after 80% saturation contained about 10 deuterium atoms with position unsaturation varying from carbon atoms 5 to 13.[31] Thus, commercially hydrogenated fat products contain isomeric products not originally present.

Three main types of hydrogenation are carried out: a) selective and b) nonselective to give shortening and margarine oils, and c) a combination of the two to give stearine-like products. Selective hydrogenations are carried out at high temperature and low pressures, such as 175°C, 5 psi of hydrogen, and 0.05% catalyst. Nonselective hydrogenations are carried out at lower temperatures and higher pressures, such as 125°C, 50 psi of hydrogen, and 0.05% catalyst. Conditions for manufacture of stearine frequently combine higher pressures and temperatures, such as 200°C, 50 psi, and 0.05% catalyst. Selectivities as measured by rates of hydrogenation for oleate, linoleate, and linolenate for nonselective conditions and a nickel catalyst are approximately 1, 7.5, and 12.5, i.e., linoleate hydrogenates 7.5 times as fast as oleate under these conditions. For selective conditions, the relative rates or selectivities are 1, 50, and 100.[32] An increase in temperature or catalyst concentration will generally increase these selectivities for linoleate or linolenate, but an increase in agitation or pressure of hydrogen will lower selectivities.[33]

Hydrogenators are primarily of three types: batch with a "dead-end" system[29] where hydrogen is not recirculated through the oil during the reduction, batch with recirculation of hydrogen, and continuous.[34] Most hydrogenators in use today are batch with the "dead-end" system. Products resulting from hydrogenation will be discussed later.

Deodorization

Most edible oil products consumed in the United States are deodorized in some type of continuous or semicontinuous equipment such as developed by Bailey[35,36] Several different approaches to deodorization equipment are possible. A review of some of these approaches was made by Zehnder and McMichael[37] and a list of manufacturers of this and other equipment available to the soybean industry is published annually in the Blue Book Issue of the *Soybean Digest,* in March of each year.

The Bailey or tray-within-a-shell design consists of a series of trays usually constructed of 18-8 stainless steel. A flow diagram of one company's six-tray deodorizer is shown in Figure 8. The oil enters through a charge pump into tray No. 1 where it is heated with steam in coils and deaerated. Proper care is required to prevent excess foaming. Deodorization is effected in trays 2 through 5 where the oil is heated with a Dowtherm system to temperatures of 260°C or higher. The oil passes through a labyrinth in each tray while steam is passed through the oil. In tray No. 6, the oil is cooled, transferred to storage, tank car, or truck through a drop tank, discharge pump, and polishing filter. Volatile materials that condense on the sides of the deodorizer flow down the inside wall of the shell to be collected at the

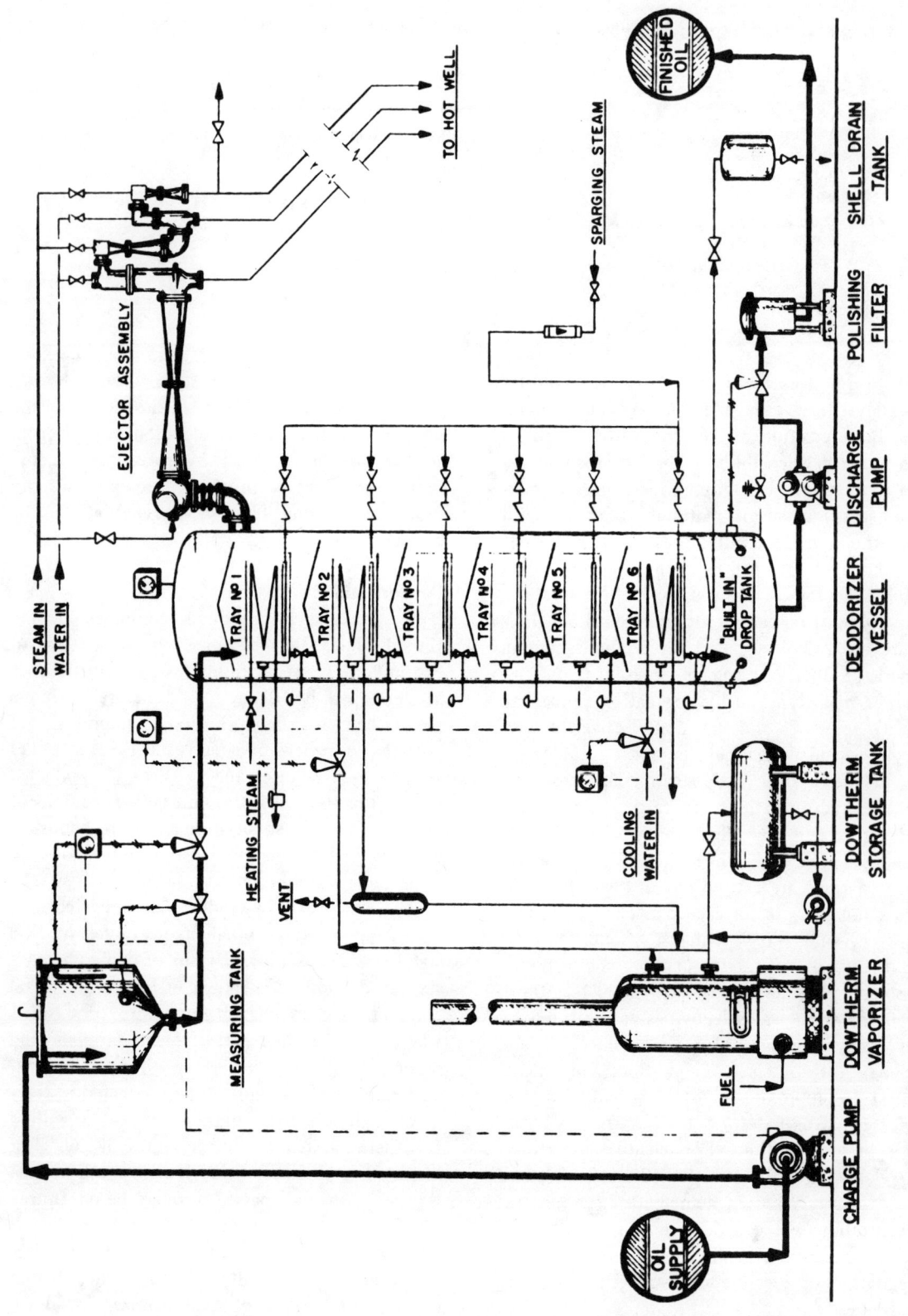

FIGURE 8. Flow diagram for six tray semi-continuous deodorizer. (From Votator Division, Chemetron Corporation. With permission.)

bottom. They are called shell drain condensates and they are periodically removed to the shell drain tank. These condensates, as well as products solidifying or collecting in the outlets of the steam-ejector pumps, are important sources of tocopherols, stigmasterol, and sitosterol for the food and pharmaceutical industry.

By the use of low pressures and high temperatures with the passage of steam through the oil, the volatile products are removed from the oil. The time required for a deodorization is directly related to the amount of steam passed through the oil, the pressure maintained by the steam-jet ejectors and the temperature. For example, increasing the temperature from 177 to 204°C will treble the rate of removal, and another increase from 204 to 232°C will again treble the rate of removal. Many deodorizers are maintained at about 6 mm pressure (mercury-absolute) by three-stage, steam-jet ejectors. Recently some operators have gone to lower pressures of 1 to 3 mm pressure (mercury). Volume of steam and the intimacy of mixing with the oil are more important than the weight of steam throughout. About 5% by weight of steam throughput is considered a normal amount.

Air entering into the oil while it is hot and oxygen in the steam are important factors in the production of a bland oil of good stability. All water used for steaming the oil needs to be thoroughly deaerated and the system operated to remove any oxygen prior to use with the oil. Direct leakage into the deodorizer can also be important, but the tray-within-a-shell design limits the effect of such leaks. Another advantage of the semicontinuous over continuous deodorizers is the ease with which stock entering the deodorizer can be changed.

Modern deodorizers of the semicontinuous type usually require about 60 minutes for the deodorization. A new design that improves the contact of the stripping steam with the oil is reported to require only 15 minutes of deodorization at 270 to 280°C or about one fourth the time of current practice.[38] This new equipment when properly operated gives substantially higher yields of tocopherols and sterols.

The effects of time and temperature of deodorization in glass on the initial quality and flavor stability of the oil indicate that commercial operations may be at or close to the top limit of desirable operation. In studies in which the temperature was varied in all glass equipment from 190 to 270°C in 20 degree increments and the time from 0.5 to 3 hours, the effect on initial quality was generally one of improvement. However, the oil after 0.5 hr at 270°C was initially superior to an oil deodorized for 2 hr at 270°C. Two hours of deodorization at 230, 250, and 270°C lowered oxidative and flavor stability.[39]

Most soybean oil is treated with phosphoric or citric acid to inactivate traces of metals, particularly iron and copper. About 0.005% citric acid is sometimes added in the cooling stage of the deodorizer. Phosphoric acid is sometimes used in a refining step, such as acid washing of alkali-refined oil, to inactivate traces of metals present. Other procedures are also used to introduce metal inactivating agents into the oil. Care is needed to avoid introduction of too much of phosphoric acid-containing products, such as phosphatides. At levels above 0.02 to 0.05%, phosphatides give rise to cucumber and watermelon flavors in the aged oil. Such oils with increased phosphoric acid or phosphatides have improved oxidative stability, but the cucumber flavors appear and they lower the flavor scores of the aged oils.[40]

From the deodorizer, the oil flows through filter to storage, tank cars, or trucks for bulk shipping, or to packaging equipment where it is bottled, canned, or put into drums. Nitrogen is used in many modern plants to blanket the oil during cooling, shipping, and packaging. The tank is filled with nitrogen and the oil passed through a pipe into the bottom of the tank. Small containers are filled with nitrogen, passed into a compartment filled with nitrogen, and loaded with an oil saturated with nitrogen and sealed before entering the air again. More than one company packages oil in dark brown bottles or cans because soybean oil is light sensitive.[41]

Stabilizers other than metal-inactivating agents are often added to soybean oil products in the cooling stage of the deodorizer. Methyl silicone in amounts of 1 to 5 ppm is added as an antifoam agent. Antioxidants include t-butyl hydroxyanisole and t-butyl hydroxytoluene and propyl gallate. The latter improves stability as measured by the active-oxygen method (AOM), and the anisole and toluene derivatives are reported to improve the stability of the oil in baked and other cooked foods. Total amounts added must not exceed 0.02%.[42]

SOYBEAN OIL PRODUCTS

A wide variety of soybean oil products is now available as consumer products or for their manufacture. Since the trend is to more and more convenience foods, soybean oil products are finding increasing outlets in them. This growth in convenience foods explains in part the increased use of soybean oil in the category known as salad and cooking oils. For many products, soybean oil is mixed with other oils or fats of increased palmitic ester content to improve plastic range, such as in shortenings, or with oils containing no linolenic ester to lower the content of this component. Since soybean oil usually contains 6 to 9% of linolenate, a mixture with cottonseed, peanut, corn, or sunflower oil will lower the percentage of this acid. The usual method for lowering linolenate content is hydrogenation.

Salad and Cooking Oils

Salad oils may be prepared directly from alkali-refined oils by deodorization, but a bleaching step usually improves color and flavor stability. Without bleaching, the oils can meet a specification used to purchase oils for shipment under PL-480 or Food-for-Peace specifications. With bleaching and proper treatment to inactivate metals, oils are suitable for use as salad dressings and mayonnaise. Taste panel and other data relating to unbleached, bleached, and citric acid-treated oils are given in Tables 7 and 8.

The more popular salad and cooking oils in the United States for years have been corn, peanut, and cottonseed oils. Some safflower, sunflower, and olive oils have also been used. In the last five to ten years, increasing amounts of partially hydrogenated winterized soybean (HWSB) oil have been sold to the consumer trade. The alkali-refined and bleached oil is hydrogenated from an iodine value of about 132 to 105 to 108 and winterized to remove fats that crystallize out at about 32°F.[44,45] Winterization is usually effected by slowly cooling the oil to about 55°F over a period of 12 hr or until the first crystals appear. The cooling brine is usually kept at about 25°F below the temperature of the oil until crystals start to form. Then over the next 18 hr the cooling rate is slowed to permit the crystals to grow and temperature to drop slowly.

When the temperature drops to about 40 to 42°F, oil is held there for a considerable length of time. Exact duration will depend on extent of winterization needed. A holding period of at least 12 hr is required to pass a cold test of 20 hr at 32°F. The oil is then filtered through plate and frame filter presses to give salad oil and solid residue, sometimes called vegetable stearine. The oil after winterization usually goes to the deodorizer for processing.

HWSB oil when properly stabilized serves both as a salad and as a cooking oil. It is an improved product when compared directly with unhydrogenated soybean oil. It contains a maximum of 3% linolenate. Flavor panels and consumer organizations have rated it below corn and cottonseed oils as a salad and cooking oil.[46]

A product derived primarily from soybean oil is used extensively for deep-fat frying of foods by some food chains and food processors. This cooking oil is prepared by hydrogenating and not winterizing to lower the iodine value to about 95 to 100 and the linolenate to as low as possible

TABLE 7

Effect of Bleaching on Flavor Stability of Soybean Oil (Both Oils with Citric Acid)

	Bleached	Unbleached
Initial flavor score	7.9	6.7
Aged Flavor score	6.4	5.2
Increase in PV on aging	1.0	3.5
PV, 8-hr AOM*	2.2	20.0

*Increase in peroxide value in active oxygen method at 8 hr. (From Cowan.[27])

TABLE 8

Effect of Citric Acid Treatment of Soybean Oil on Flavor and Oxidative Stability

	No citric acid	Citric acid (0.01%)
Initial flavor score*	8.3	8.1
Aged flavor score		
3 days at 60°C	4.0	7.5
6 weeks at room temperature	5.0	8.0
12 weeks at room temperature	3.8	7.5
18 weeks at room temperature	2.7	6.0
Increase in PV† at 18 weeks	4.0	0.3

*Scores on a basis of 1 to 10.
†PV = peroxide value.

(From Dutton et al.[43])

without loss of fluidity at room temperature. This product will be cloudy at room temperature but will still pour readily. Since most shortenings do become fluid at elevated temperatures, these products approach shortenings in some of their properties. Indeed, some liquid as well as "solid" shortenings serve as cooking oils. Specifications for fully refined soybean salad oil as suggested by the National Soybean Processors Association (NSPA)[47] and for the purchase of soybean oil for overseas shipment by Agricultural Stabilization and Conservation Service (ASCS)[48] and for HWSB oil as suggested by ASCS for purchase of HWSB oil are shown in Table 9.

In Table 10, a comparison of the compositions of unhydrogenated soybean oil, three HWSB oils, and a liquid shortening based on HWSB oil is shown.[44,49,50] For those people interested, polyunsaturated content and polyunsaturated/saturated fatty acid ratios are also shown.

Shortenings and Margarine Oils

Since soybean oil products will normally contain only about 11% of palmitic acid, their effectiveness as a shortening has limitations. Shortening power is readily obtained by use of mixtures that give wide ranges in melting points and solid contents. Soybean oil products are substantially improved by mixing with cottonseed oil products that have 20 to 23% palmitic, lard that has 20 to 28%, and tallow that has 24 to 37%. Palmitic esters generally melt lower than stearate esters and wider plastic ranges are more readily achieved with the mixtures of soybean oil with these other fats than with soybean oil alone. Extent of hydrogenation will depend on the oil or fat used and the widest plastic ranges are obtained with mixtures of oils and hydrogenated fats. Many shortenings are formulated to be not too hard at 50°F or too fluid or soft at 90 to 110°F. General-purpose and high-stability shortenings have similar characteristics, but a greater percentage of liquid triglycerides will be found in the general-purpose shortenings.

During the 1960's, the composition of the all-purpose plastic shortenings changed to include a higher percentage of polyunsaturates. Compositions for samples of two types of shortening are shown in Table 11. The solid fat indices of these shortenings, as well as an all-purpose shortening from a blend of animal and vegetable fats and high-stability shortenings made of blends of hydro-

TABLE 9

Specifications for Soybean Oil

	Fully refined salad oil ASCS	Fully refined salad oil NSPA	HWSB salad and cooking oil ASCS
Appearance	Clear and brilliant at 70 to 85°F		
settlings	No settlings or foreign matter of any kind		
Odor and flavor	Bland and free of rancid and certain other odors and flavors		
Moisture and volatiles (min), %		0.1	0.06
Flash point	550°F		550°F
Free fatty acid, % as oleic (min)	0.05	0.05	0.05
Refined bleach color			
Lovibond red	2.0	2.0	2.0
Lovibond yellow		20.0	20.0
Peroxide value (meq/kg) max	0.5	0.6	2.0
Peroxide value (meq/kg) in			
8-hr AOM test	35.0	8.0	35.0
Linolenic (max), %		3.0*	
		3.5†	
Cold test (hr)		5.5	5.5

*3.0 by alkali isomerization procedure.
†3.5 by gas liquid chromatographic method.

(From NSPA,[47] ASCS.[48])

TABLE 10

Composition of Representative Samples of Soybean Salad and Cooking Oils

Analysis	Unhydrogentated salad oil		HWSB oils		Liquid shortening (cooking oil)
Iodine value	131	115	109	109	100
Palmitic, %	10	8	10	10	11
Stearic, %	5	3	4	3	10
Oleic (monene), %	26	47	47	46	45
cis			39		35
Linoleic (diene), %	52	37	36	39	30
cis,cis			29		22
Linolenic (triene), %	8	4	3	2	2
trans, Isomer (elaidic), %	–	12	15		18
Cold test	Pass		Pass		
P/S ratio*	4		2.3		1.1

*Polyunsaturated/saturated fatty acid.

(From McOsker,[44] Jones et al.,[49] Scholfield et al.,[50] Cowan.[63])

TABLE 11

Composition of Representative Samples of Plastic Shortening of High and Low Polyunsaturate Content

	High polunsaturate content		Low polyunsaturate content	
Iodine value	95	86	76	76
Palmitic, %	13	17	13	15
Stearic, %	11	11	13	10
Oleic (monene), %	45	41	62	62
cis, %	34		44	
Linoleic (diene), %	27	27	13	12
cis, cis %	23		9	
Linolenic (triene), %	3	2	0	1.0
trans, Isomer, %	14	16	21	29
P/S ratio*	1.0	∼1.0	0.4	∼0.3

*Polyunsaturated/saturated fatty acid.

(From McOsker,[44] Jones et al.,[49] and Scholfield et al.[50])

genated vegetable oils and hydrogenated animal and vegetable oils, are given in Table 12.

Soybean oil is used in large quantities in margarines that are prepared with a blend of hydrogenated oils. Usually, the oils that are hydrogenated are a blend of soybean and cottonseed with soybean percentage as high as 80. The harder component is prepared under nonselective conditions and the other softer component under selective conditions. Table 13 gives data on such margarine oils, including a tub margarine. The latter became popular in the late 1960's to supply products that were more spreadable on immediate removal from the refrigerator and contained a higher polyunsaturated fatty acid content than other margarines. The fat components of these margarines are usually a blend of a liquid oil and a hydrogenated solid fat. Hydrogenated soybean and cottonseed oils similar in nature to the harder component in Table 13 are a common ingredient in such margarines. The liquid component is usually a refined bleached deodorized corn, safflower, sunflower, or cottonseed oil. Unhydrogenated soybean oil is less often a component of the liquid oil, but HWSB oil may increase the amount of soybean oil used in tub margarine.

In the manufacture of both margarines and shortenings, the melted blend of oils is processed through chilling machines to control the crystal structure of the product. Over 90% of all margarine and shortening is now processed through chilling equipment similar to that developed by Joyner.[52] This equipment is known as a Votator. For shortening, a mixture of hydrogenated oils is pumped into a small closed system where the fat is continuously solidified on externally refrigerated cylinders and scraped therefrom. The fluid fat is supercooled to 60 to 65°F and small crystals form. The supercooled fluid mixture passes into a second unit or worker unit to continue the growth of the small crystals without cooling. The shortening is packaged and allowed to temper at about 80°F for one to three days to give proper crystal structure.

TABLE 12

Properties of All-Purpose and High-Stability Shortenings Made from All-Hydrogenated Vegetable Oils and Blends with Animal Fats

	High-stability shortening		All-purpose shortening		
	All-vegetable	Blend with animal fat	All-vegetable polyunsaturates		Blend with animal fats
Iodine value	69	58	74	91	60
Linoleic %	1.5	3.5	8	30	8
Solid fat index (% solid at temperature indicated)					
10°C	40	39	28	16	29
21.1°C	27	26	22	15	22
26.7°C	22	22	20	14	20
33.3°C	11	14	16	13	17
40°C	5	9	12	11	13
Melting point °F	109	115	123	123	

(From Mattil[51] and unpublished data.)

TABLE 13

Some Analytical Data for Margarine Oils

Analysis	Harder component	Softer component	Blend low in poly.*	Blend high in poly.*
Iodine value	55	86	79	95
Melting point °C	44	30	38	30
Solid fat index				
(% solid at temperature indicated)				
10°C	66	18	27	21
21.1°C	59	7	16	14
26.7°C	57	2	12	1.5–3.5
33.3°C	43	0	3.5	0
37.8°C	27	0	0	0

*Polyunsaturated fatty acids; a blend high in polyunsaturates may have as much as 30% or more depending on the liquid oil used in the blend.

(From Mattil[29] and unpublished data.)

This tempering usually improves consistency and creaming characteristics. For margarine the equipment is similar, but operations are different and vary somewhat with the type of margarine.[53] The milk ingredient, such as ripened skim milk, is emulsified with the margarine oil and other minor ingredients, such as vitamins, salt, and lecithin, and pumped into a series of externally cooled cylinders where the solidified fat is continuously scraped from the cooling surfaces. For cube margarines, the emulsion is cooled to 45 to 55°F by passing through three refrigerated scrapers (first unit) into a second or holding unit. Unlike the shortening or "tub" margarine operations, the emulsions are not stirred in this second unit. These static or second units are usually operated in pairs to provide one holding unit while the other is furnishing material for cubing and packaging. For whipped margarine, gas is injected into one of the refrigerated scrapers and a worker-type crystallizer is used just before forming the cubes. For tub margarine, less refrigerated scraper capacity is needed but a larger worker-type crystallizer is used before filling the cartons.

There are also many specialty products that contain hydrogenated soybean oil in varying amounts, such as fats for frozen desserts or mellorine, for cookie shortenings, confections, icings, ice cream coatings, whipped toppings, and coffee whiteners to mention a few.

Flavor Stability of Soybean Oil

As mentioned previously under deodorization, citric acid improves the flavor and oxidative stability of soybean oil (see Table 8). Citric acid works this apparent magic by inactivating trace amounts of metals in the oil.[54] Failure of citric acid to improve soybean oil treated with a nonacidic sorbitol suggests that citric acid is only a metal-inactivating agent and not a synergist for antioxidants present.[55] Recognition of the value of citric acid and its role as an inactivating agent for metals was slow to come because often less than 0.3 ppm of copper or iron was involved and the discoverer and users of citric acid kept it an industrial secret in Europe prior to World War II. W. H. Goss, a member of the Technical Industrial Intelligence Committee, listed the use of citric acid as one of a number of procedures that improved the stability of soybean oil.[56] Subsequent laboratory[43] and commercial studies[57] confirmed the special value of citric acid for soybean oil. In addition, a number of nitrogen-, sulfur-, and other oxygen-containing acids were found or synthesized that proved to be equally effective, but citric and phosphoric acids are still the primary inactivators used in the industry.

Although iron impurities accelerate the oxidation of other oils, they do have a greater effect on soybean than on cottonseed oil at low levels such as 0.3 ppm of added iron.[58] In Table 14, note that 0.3 ppm of added iron impurity had little effect on either the initial or aged flavor score of cottonseed oil, but the same amount of iron had a significant effect on the flavor scores of the initial and aged soybean oil. In the 8-hr AOM test, the added iron raised the peroxide value of the soybean oil about 2.5 times as much as it raised the peroxide value of cottonseed oil.

The off-flavors from refined, bleached, and deodorized soybean oil when the oil is aged in a variety of ways differ in part from the off-flavors derived from cottonseed, peanut, or sunflower oil. The major precursor of these flavors that differentiates soybean oil from the other oils is the linolenate ester.[59] Dutton and co-workers established that linolenate molecularly incorporated into cottonseed oil imparted to it the characteristics of flavor instability usually associated with soybean oil.[60] Some reports indicate that hydrocarbons in the unsaponifiable fraction may be a secondary precursor.[61,62] Further proof for linolenate's role in flavor reversion is found in an improvement in flavor stability that results when the oil is hydrogenated to remove linolenate. A lowering of the linolenate from 6 to 9% to 3% found usually in HWSB oil results in considerable improvement. Further improvement results when the linolenate content is lowered to less than 1%, preferably to where it is no longer present. Table

TABLE 14

Effect of 0.3 ppm of Iron Added During Deodorization on the Flavor Score and Oxidative Stability of Soybean and Cottonseed Oils

	Cottonseed oil		Soybean oil	
	0.0 ppm added	0.3 ppm added	0.0 ppm added	0.3 ppm added
Initial flavor score	8.4	8.4	7.8	6.0
Aged flavor score	5.1	4.9	5.1	3.4
Peroxide value, 8-hr AOM	17	29	14	44

(From Evans et al.[58])

15 shows data on a copper-reduced oil of 0.0% linolenate compared by taste-panel techniques with an unhydrogenated oil of high quality. Note that both oils had excellent stability at 60°C for four days, but the 0.0% linolenate was significantly higher in flavor score after eight days at 60°C and at the same level of oxidation as measured by peroxide values.[63]

Perhaps the quickest way to show differences between soybean oils is to heat them to an elevated temperature, cool to 55°C, and taste. The 0.0% linolenate oil was significantly higher in flavor score when heated to 150 or 200°C and at about the same levels of peroxide values. An examination of the flavor responses of soybean oils of different linolenate content heated to 200°C and cooled to 55°C for tasting gave the results shown in Table 16.

One of the major comments of European processors about soybean oil for frying is the odor in the room. The differences in odor responses between a soybean oil containing no linolenate and one containing 7.8% are also shown in Table 16.

HWSB oil usually contains about 3.0% linolenate and it represents an improvement over unhydrogenated soybean oil by many test procedures such as aging at 60°C, heating to elevated temperatures, and testing of room odors.[63] With most HWSB oils now available, an antifoam agent is used as well as antioxidants, metal-inactivating agents, and inhibitors to prevent crystallization of any fatty glycerides from the oil. The antifoam agent, methyl silicone, is particularly helpful in improving quality of the oil for frying. Its use apparently reduces the amount of oxygen incorporated into the oil during frying. Studies with HWSB oil (3% linolenate), copper-reduced oil (0.0% Linolenate), and a high quality cottonseed oil show that copper-reduced soybean oil is probably intermediate in flavor quality between the other oils. Data on the oils obtained with the heat and room odor test are shown in Table 17. The failure of copper-reduced oil to score as high as cottonseed oil in some of these tests may be the effect of isolinoleic acids present and the odors that may result from their oxidation.

The reports as to what the compounds are that cause flavors in soybean oil have shown that a large number of compounds are present. Leading candidates for the principal culprits are: 2-pentyl furan reported by Smouse and Chang;[64] ethyl vinyl ketone, pentanal, 4-cis-heptenal, and diacetyl reported by Hill and Hammond[65] and by Seals and Hammond;[66] 3-cis-hexanal for the beany flavor suggested by Hoffmann;[67] and 6-trans-nonenal as the flavor characteristic of hydrogenated soybean oil.[68]

These off-flavors are unquestionably produced by oxidation as shown by the improvement in shelf life at 100°F of HWSB oil when packed under nitrogen. For example HWSB oil that was packed in air dropped in flavor score in 10 weeks at 100°F from 8.0 to 5.0, whereas HWSB oil packed under

TABLE 15

Comparison of 0.0% Linolenate Oil with Original Unhydrogenated Soybean Oil Showing Flavor Scores and Significance

Condition	Linolenate 0%[a]	s Sig[b]	Linolenate 7.8%[a] (original)
Initial flavor	7.7(0.0)	+	7.8(0.0)
Aged, 60°C, 4 days	6.5(0.8)	+	6.2(0.7)
Aged, 60°C, 8 days	5.7(6.2)	**	4.2(6.2)
Heat test 150°C	5.3(0.7)	*	4.2(0.9)
Heat test 200°C	3.8(3.6)	**	2.2(2.8)
Peroxide value, 8hr, AOM	5.0		5.8

[a]Peroxide values are given in parentheses after flavor scores.
[b]+ = no significant difference at 5% level.
* = significant difference at 5% level.
** = significant difference at 1% level.

(From Cowan et al.[63])

TABLE 16

Flavor or Odor Intensity Values* of Soybean Oils in Heat and Room Odor Tests at Different Levels of Linolenate

	Soybean oils			
% Linolenate	0.0	1.3	2.0	7.8
Oils heated to 200°C, cooled, and tasted – Flavor Intensities				
Rancid	1.2	1.2	1.4	1.4
Painty	0.3	0.7	1.0	1.6
Fishy	0.0	0.1	0.0	0.8
Oils heated to 193°C in fry pan – Odor Intensities				
Hot oil	0.7			0.2
Rancid	0.1			0.6
Fishy	0.2			1.4

*Flavor or odor intensity value = (weak responses + 2 × medium responses + 3 × strong responses) divided by the number of tasters.

(From Cowan et al.[63])

nitrogen dropped from 8.0 to about 7.3 in 10 weeks and to only about 7.0 in 26 weeks.[69]

Soybean Lecithin - Products and Use

The crude dried lecithin from the degumming operation is an important raw material for the manufacture of a variety of lecithin products for the industry.[70] Soybeans contain 1.5 to 3.0% phosphatides and a considerable portion of this amount is extracted from the flakes by hexane. With over 700 million bushels of soybean processed each year, there are large quantities of phosphatides available and more and more of them are finding outlets in industrial nonfood as well as food uses. Phosphatides have been primarily a minor ingredient for their surface-active properties as emulsifiers, foam stabilizers, suspending agents, release agents, nutritive supplements, wetting agents, slip agents for plastics, and antispattering agents.[71]

The crude dried product from degumming is usually bleached with hydrogen peroxide to lighten the color and is often mixed with 2 to 5% of soybean fatty acids to improve fluidity. Some of the characteristics of these products are shown in Table 18. All grades can be obtained as unbleached, bleached, or double bleached with colors on the Gardner scale of 10, 7, and 4, respectively. The acetone-soluble part of these phosphatides is primarily oil with some fatty acids.

The oil can be separated from the phosphatides by extraction with acetone to give a dry granular product. One grade of this product is available at drug and dietary food stores under the tradename of RG Lecithin. In addition, the oil-free lecithin can be further separated with alcohols, such as isopropyl alcohol, into fractions with large amounts of phosphatidyl choline (soluble) and inositol phosphatides (insoluble), respectively. Table 19 summarizes properties of these products.

Although the oil-containing lecithins are used in most products, the oil-free phosphatides and their derived fractions have special properties that make them more desirable for certain end-products. It is not our purpose here to review all of the uses for the phosphatides but to mention some representatives examples. The emulsifying properties of the phosphatides find extensive use in baked goods, cake mixes, and instant foods, where variable amounts are used depending on the product about 0.1 to 0.3% in bread based on the flour used, 0.5 to 1.0% in pies based on shortening used, 1.0 to 3.0% in cake mixes based on shortening, and 0.5 to 3.0% in instant foods to improve their dispersibility. Lecithin improves the quality of many candies including nougats, caramels, and taffies. When added to certain clear, brittle high-sugar candies, the phosphatides improve the bril-

TABLE 17

Comparison of HWSB Oil with Copper-Reduced Oil and Cottonseed Oil

Condition	HWSB oil	Copper reduced	Cottonseed	Sig*
Initial flavor score	7.4	7.4	8.1	+
Heat test at 200°C				
Odor score		3.9	4.7	+
	4.4		5.6	*
	3.9	5.1		**
Flavor score		3.1	3.9	**
	3.2		4.6	*
	2.8	3.9		**
Room odor score		5.9	6.4	+
	6.0		7.1	*
	5.7	6.1		+
Odor Intensity Values†				
Room Odor Test				
Hot oil	0.5	0.7	0.4	
Rancid	0.6	0.6	0.5	
Fishy	0.6	0.2	0.2	

*See Table 15 for definition.
†See Table 16 for definition.

(From Cowan et al.[63])

TABLE 18

Characteristics of Some Different Grades of Bleached Lecithin Products

Consistency	Plastic	Fluid	Fluid	Fluid
Acetone insoluble, %	67–72	62–64	54–60	62–64
Moisture, %	1.1	0.75	0.75	0.75
Benzene insoluble (max), %	0.1	0.1	0.1	0.05
Acid value	25	32	40	32
Centipoises at 80°F	–	15,000	7,000	15,000

(From Technical Sales Manual.[71])

liance and clarity of the candies.

The alcohol-insoluble fraction imparts improved antispattering characteristics to margarine and finds use in pan greases, peanut butter, and table spreads. The alcohol-soluble fraction gives more stable oil-in-water emulsions whereas the alcohol-insoluble fraction gives more stable water-in-oil emulsions. A specially prepared nonpyrogenic alcohol-soluble fraction is useful in preparation of fat emulsion for intravenous feeding.

The phosphatides are frequently put in different carriers to improve color or flavor in the final product. Some examples include incorporation into cocoa butter for chocolate coatings; into shortenings for baked goods; into margarine oils for margarine; and onto flours for cake and other mixes.

Lecithin can be hydroxylated with hydrogen peroxide and lactic or other acids to improve hydrophilic properties such as moisture retention and improved efficiency for oil-in-water emulsions.[72] These properties make it more useful in a variety of baked products.

FOOD USES OF SOYBEAN PROTEINS

Physical and Chemical Properties

Before discussing the various processed forms of soy proteins—flours, concentrates, and isolates—used in the food industry, it is desirable to review physical and chemical properties of the proteins in these products. Many proteins including those of soybeans are sensitive to physical and chemical treatments that may have little or no effect on other food components. Moist heat and extremes of pH are examples of two processing conditions that can dramatically alter physical properties of soy proteins such as solubility, molecular weight, and viscosity.

Solubility as Function of pH

The majority of soy proteins are globulins. This class of proteins is insoluble in water in the region of their isoelectric points but will dissolve in the isoelectric state when salts such as sodium or calcium chloride are added. If the pH is above or below the isoelectric point, a globulin will also dissolve in aqueous solutions in the absence of salts. The effect of pH on solubility of soybean proteins is clearly illustrated in Figure 9. These results were obtained by stirring undenatured, defatted meal in water, adding acid or alkali to vary pH, centrifuging, and analyzing the extracts

TABLE 19

Composition and Properties of Oil-Free Phosphatides and Their Alcohol-Soluble and Insoluble Fractions

Property	Oil-free lecithin	Phosphatidyl choline fraction	Inositol phosphatide fraction
Solubility in			
Oil	Soluble	Soluble	Soluble
Water	Dispersible	Dispersible	Dispersible
Alcohol	Dispersible	Soluble	Insoluble
Phosphatidyl choline, %	30	60	4
Phosphatidyl cephalin, %	30	30	28
Inositol phosphatides, %	32	2	55
Oil, %	3	4	4
Miscellaneous, %	5–6	4–5	8–9
Type of emulsions	O/W or W/O*	O/W	W/O

*O/W = oil-in-water; W/O = water-in-oil.
(From Technical Sales Manual.[71])

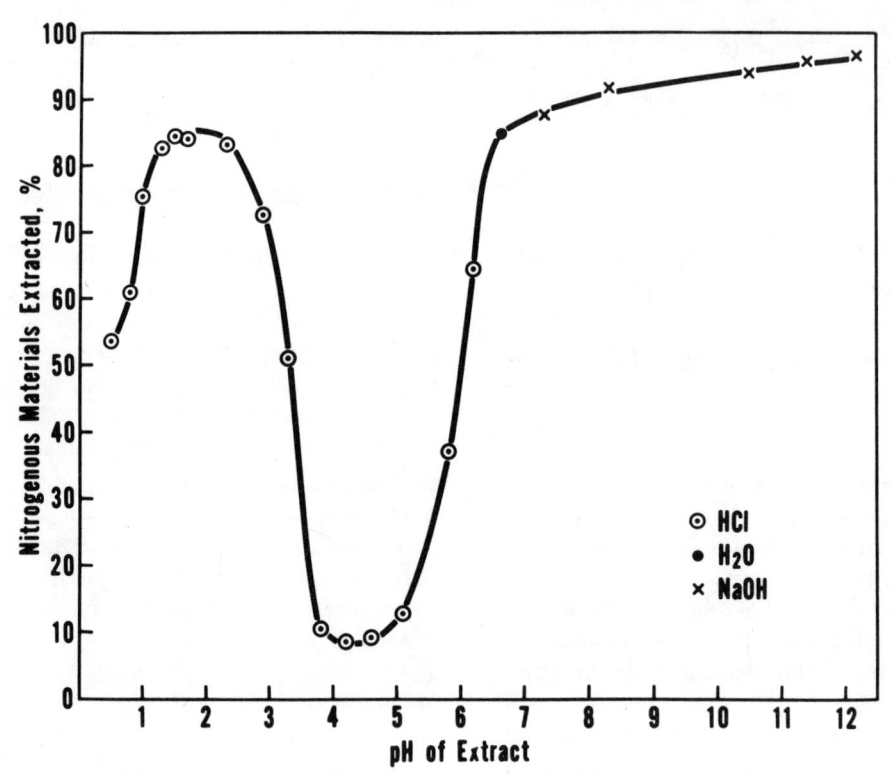

FIGURE 9. Extractability of proteins in defatted soybean meal as a function of pH. (Data from Smith and Circle.[73])

by the Kjeldahl method. A suspension of defatted meal in water has a pH of about 6.5 and about 85% of the nitrogenous components (mainly proteins) dissolves. Addition of alkali increases extractability by another 5 to 10%; but if acid is added, protein solubility decreases abruptly and reaches a minimum at pH 4.2-4.6, the isoelectric region. Insolubility of the globulins at this pH is used to prepare protein isolates as described later (Forms of Soy Proteins). Further addition of acid causes resolubilization of the proteins below the isoelectric point. The pH-solubility relationship of soybean proteins is similar to the pH dependency of solubility for casein which has an isoelectric point of 4.6. Consequently, soy isolates can often substitute for casein and caseinates in food products. Such substitutions, however, require careful evaluation because the two proteins also differ in many respects.

Soy globulins can be modified with pepsin to make them soluble in the isoelectric pH region. Proteins of this type, however, are extensively hydrolyzed and have much lower molecular weights than the unmodified globulins. Pepsin hydrolyzates are used mainly for their foaming properties and are incorporated into foods which are slightly acidic, such as candies, where the unmodified proteins are poorly soluble.[74]

Molecular Size

The range of molecular sizes of soybean proteins is readily demonstrated by ultracentrifugation.[75, 76] The water-extractable proteins of defatted meal give a typical ultracentrifuge pattern in which four major fractions are resolved (Figure 10). The fractions are designated 2, 7, 11, and 15S based on their sedimentation rates. Approximate amounts of the four fractions are given in Table 20. Several of the ultracentrifugal fractions are mixtures, and proteins that have been isolated from the four ultracentrifugal fractions are listed in Table 20. Additional proteins are likely to be found on further fractionation. The 2S fraction, representing about 20% of the total protein, contains several trypsin inhibitors, cytochrome c, and unidentified proteins.

The 7S fraction constitutes a little over a third of the total protein and contains four different types of proteins: four hemagglutinins,[86] two or more lipoxygenases,[87,88] β-amylase, and a protein

TABLE 20

Approximate Amounts and Components of Ultracentrifuge Fractions of Water-Extractable Soybean Proteins

Fraction	Percent of total*	Components	M.W.	Reference
2S	22	Trypsin inhibitors	8,000–21,500	78, 79
		Cytochrome c	12,000	80
7S	37	Hemagglutinins	110,000	81
		Lipoxygenases	102,000	82
		β-Amylase	61,700	83
		7S Globulin	180,000–210,000	84
11S	31	11S Globulin	350,000	76
15S	11	–	600,000	

*Data from Wolf et al.[85]

called 7S globulin. The lipoxygenases are of special interest to the food technologist since they oxidize polyunsaturated acids which in turn undergo further reactions to form undesirable flavors. The 7S globulin is a glycoprotein[89] and makes up over one half of the total 7S fraction.[90] The 11S fraction accounts for an additional one third of the total soybean protein and thus far only a single protein, the 11S globulin, has been isolated. The 15S fraction makes up the remaining one tenth of the protein. This fraction has not been isolated and studied but appears to have a molecular weight of over half a million on the basis of its sedimentation rate. Often one observes ill-defined peaks sedimenting ahead of the 15S fraction; these proteins are designated as the >15S fraction and may be aggregates of the 15S fraction. Some workers believe that the 15S and >15S fractions are aggregates of the 11S fraction.[91] The major conclusions to be drawn from Table 20 are that soybean proteins comprise a complex mixture and that the bulk of the proteins have molecular weights of 100,000 or more.

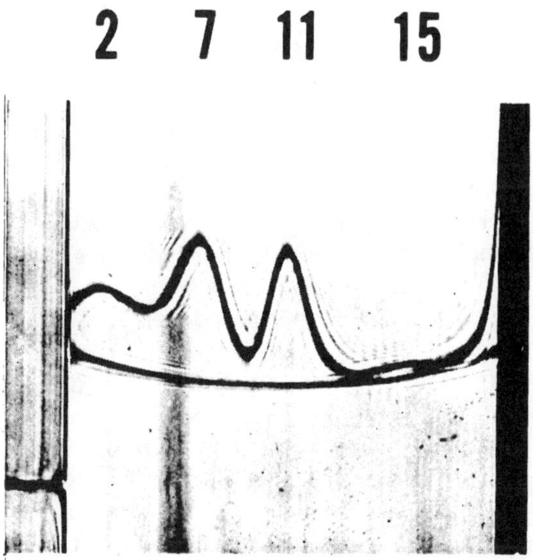

The wide distribution of molecular sizes among the water-extractable soybean proteins is also observed by gel filtration (Figure 11A). Seven protein fractions are separated as compared to only four in the ultracentrifuge. Fraction 1 was attributed to turbidity but appears to contain ribonucleic acid.[93] Figure 11B shows ultracentrifugal compositions for the other fractions. Most of the 11S protein ($s_{20,w}$=12.1S) eluted in fraction 3 but also contained a 7S protein. The remaining proteins of the 7S group eluted in fractions 4 to 6 while the 2S proteins occurred in fractions 7 and 8. The absorbance in fractions 9 to 11 appeared to be caused by nonprotein contaminants.

Additional evidence for complexity of soybean proteins has been obtained by hydroxylapatite chromatography,[94] starch gel electrophoresis,[95,96] and immunoelectrophoresis.[97]

Reactions of the 7S and 11S Globulins

The 7S and 11S globulins, which are the two major soybean proteins, have been purified and characterized. Although they are deserving of much more study, several interesting properties are

FIGURE 10. Ultracentrifuge pattern for water-extractable soybean proteins. Solvent is pH 7.6, 0.5 ionic strength buffer containing 0.01 M mercaptoethanol. Numbers across top of pattern are sedimentation coefficients in Svedberg units. (Taken from Wolf.[77])

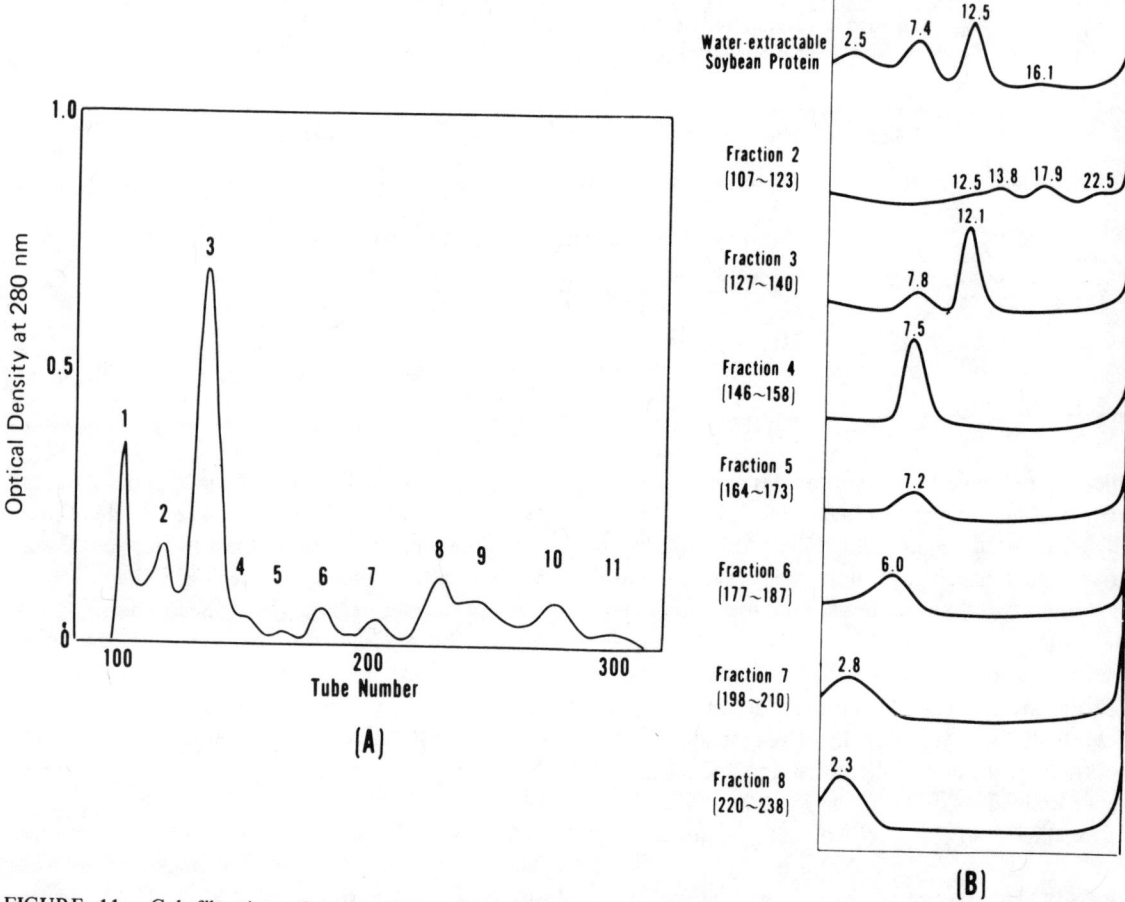

FIGURE 11. Gel filtration of water-extractable soybean proteins on Sephadex G 200; (A) Elution diagram; (B) Ultracentrifuge patterns for fractions obtained. Numbers in parentheses are tube numbers for each fraction. Sedimentation coefficients are given above peaks. (From Hasegawa, K., Kusano, T., and Mitsuda, H., *Agric. Biol. Chem.*, 27, 878, 1963. With permission.)

known. Both proteins form disulfide-linked polymers which occur in water extracts of defatted meal.[90] Further polymerization occurs during isolation procedures involving precipitation of the proteins.[98,99] The disulfide polymers cause insolubility of isolates,[99] as discussed later, and are responsible for turbidity and for increasing viscosity of solutions of the proteins.[100] The disulfide-linked proteins are easily depolymerized by mercaptoethanol, cysteine, or sodium sulfite.[98,101]

A second property exhibited by the 7S and 11S globulins is related to their structure; both are made up of subunits. This property also emphasizes differences between the two proteins. When the 11S protein is dialyzed against 0.01 ionic strength buffer at pH 7.6 its quaternary structure breaks down; 7S (different than 7S globulin) and 2-3S forms appear.[102] Under these same conditions the 7S globulin exists as a dimer. Eight glycine, two phenylalanine, and two leucine (or isoleucine) amino terminal residues are reported per mole of 11S protein.[103] These results indicate a minimum of 12 polypeptide chains and therefore 12 subunits per molecule if there are no disulfide crosslinks between the polypeptide chains. However, only six different subunits separate on isoelectric focusing of the protein in urea-mercaptoethanol solution. A dimeric structure of two identical monomers made up of six subunits each is proposed for the 11S molecule.[104] Other conditions causing disruption of the quaternary structure of the 11S protein are high and low pH,[91,105] high concentrations of urea or detergents,[102,104] mixtures of phenol-acetic acid-mercaptoethanol-urea,[106] and temperatures above 80°C.[91,107]

Nine amino-terminal residues are reported per

mole of 7S globulin, and presumably there are nine subunits in the quaternary structure.[84] At low salt concentrations in acid solutions the 7S globulin is converted into two slower sedimenting species with coefficients of 2S and 5S. Conversion of 7S into 2S and 5S forms at low pH is inhibited by salts and dialysis of the protein to pH 7.6, 0.5 ionic strength, reverses the changes in sedimentation properties. At high pH (0.01 N sodium hydroxide) the 7S protein changes to a form with an $s_{20,w}$ value of only 0.4S.[108] Formation of the slow-sedimenting protein forms is indicative of a subunit structure, and molecular-weight determinations have confirmed this interpretation.[109]

Spinning of protein isolates into fibers for textured foods is an example of a food system in which the quaternary structures of the 7S and 11S globulins are disrupted. The isolates are dissolved in sodium hydroxide at pH 12 to form the spinning dope. The high pH causes an increase in viscosity and conversion of all the proteins into a 3S form. The spinning dope is then extruded through spinnerets into an acid-salt bath which coagulates the proteins. Disulfide bonds and noncovalent interactions between the protein subunits stabilize the three-dimensional structure of the fibers.[110]

Both globulins are sensitive to changes in their ionic environment in aqueous systems. When ionic strength is changed, the proteins undergo association-dissociation reactions. The 7S globulin shows this phenomenon most clearly. At pH 7.6, 0.5 ionic strength, the 7S globulin has a molecular weight of 180,000 to 210,000 (monomer form), but when the ionic strength is lowered to 0.1 the protein sediments as a 9S component and has a molecular weight of 370,000 as a result of dimerization.[84,108] The 11S globulin is likewise converted to a faster sedimenting form when ionic strength is changed from 0.5 to 0.1, but the degree of association is small.[75] Both proteins undergo association-dissociation reversibly. In most food systems salt concentrations are sufficiently low to favor the associated forms of the two proteins.

A recent study indicates that the 7S and 11S globulins have significantly different properties when incorporated into foods. Tofu (bean curd) made from crude 11S protein was firmer in texture than tofu made from a crude 7S preparation.[111] These results suggest that the two proteins may differ significantly in their functional properties and that further comparative studies of these proteins should be made.

Solubility of Isolates

As discussed in more detail later (Forms of Soy Proteins), isolates are prepared from aqueous extracts of defatted flakes or flour by adjusting the pH to the isoelectric region of the proteins, which treatment precipitates the globulins. This process reversibly and irreversibly modifies solubility of the proteins. Although initially soluble in pH 7.6, 0.5 ionic strength buffer, the proteins are no longer completely soluble after isoelectric precipitation at pH 4.5. Laboratory preparations of isolates vary considerably in their solubility, but addition of 0.01 M mercaptoethanol increases solubilities. The 7S and 11S fractions account for this behavior.[99] Portions of these proteins exist as disulfide-linked polymers in soybean flakes and flour and dissolve in water in the initial extraction.[90] Further aggregation occurs during isoelectric precipitation, thereby insolubilizing the proteins. This form of insolubilization is reversed by mercaptoethanol and other disulfide-cleaving reagents.

Solubilities of commercial isolates vary even more than laboratory samples, and disulfide polymers also occur in these materials.[99] Isolates from different manufacturers are similar in their chemical compositions but dissimilar in physical properties because of processing variations. All sources of supply should therefore be considered when isolates are evaluated for a given food application.

The reactions involved in irreversible insolubilization of soy globulins (i.e., formation of fraction remaining insoluble in buffer containing 0.01 M mercaptoethanol) by acid precipitation are still unknown. Portions of the 2S, 7S, 15S, and >15S fractions undergo these reactions. The 7S fraction shows a very rapid irreversible insolubilization on adjustment to pH 4.5 followed by a slower decrease in solubility with time. Two different 7S proteins or two different reactions with a single 7S component appear to be involved.[112]

Acid-precipitated soybean globulins contain significant amounts of phosphorus which is largely phytate.[113] Phytate is a highly charged anion that interacts with the globulins when they are precipitated with acid. Removal of phytate from acid-precipitated globulins shifts the isoelectric point from pH 4.5 (Figure 9) to pH 5.0. The proteins are more soluble in the region below pH 4 as a result of this shift in isoelectric point.[114] A better understanding of the effects of phytate on the physical properties of isolates would be desirable.

Denaturation

Soybean proteins are sensitive to most of the ordinary conditions known to denature proteins. Of primary interest to the food technologist are the effects of moist heat and extremes of pH, hence only these are discussed here.

Heat Denaturation—Since most foods are heated during one or more stages of processing, this form of denaturation is most commonly encountered but is also least understood. The best known effect of moist heat on soybean protein is insolubilization in water or salt solutions. Rapidity of protein insolubilization on steaming soybean flakes is illustrated in Figure 12. Solubility of the protein decreases from an initial value of 80% to 20 to 25% after heating for only 10 min. Defatting the flakes before steaming has no significant effect on rate of insolubilization.[115]

Because the proteins are so readily insolubilized by moist heat, protein solubility measurements have been used to determine extent of heat treatment given to soy products, particularly flakes, grits, and flours. Various empirical methods have been proposed.[116] Currently, methods have been narrowed down to two—the so-called "solubility" and "dispersibility" methods.[117] Hopefully, only a single method will be adopted in the future to provide for uniformity in the industry and elimination of confusion from multiple procedures and terminologies.

Both methods involve extracting the soybean product with water and analyzing the extracts by the Kjeldahl method. The methods differ primarily in the conditions for preparing the extracts and hence in the results obtained. In the "solubility" or "slow-stir" method, the extraction is made by stirring slowly for 2 hr and then centrifuging. In contrast, the "dispersibility" or "fast-stir" procedure uses only a 10-min extraction with a high-speed mixer equipped with cutting blades.

The literature contains a variety of terms to denote the results obtained by the two procedures. Because these terms are often confusing and sometimes poorly understood, we have summarized them in Table 21 along with their method of calculation from the results of the Kjeldahl analysis. Inspection of Table 21 shows that NSI is analogous to PDI, and WSP is the counterpart of WDP. Differences in numerical values for the analogous terms arise from the variations in extraction conditions between the two methods. It is also apparent that NSI and PSI are numerically identical since substitution of WSN x 6.25 for WSP leads to the expression for NSI.

The terms most frequently encountered in the recent literature are NSI and PDI. For a given sample, the PDI will generally be higher than the NSI value, apparently because of a greater shearing action during extraction than in the NSI procedure. Experimental variables that influence NSI and PDI values are discussed elsewhere.[116,118]

In laboratory studies (Figure 12) and in commercial practice NSI and PDI of soybean flakes and flours decrease with increased time of steaming and then level off. However, if soy flour is heated with an equal weight of water at 110 or 120°C, protein solubility rapidly reaches a minimum and then increases again.[119] The protein

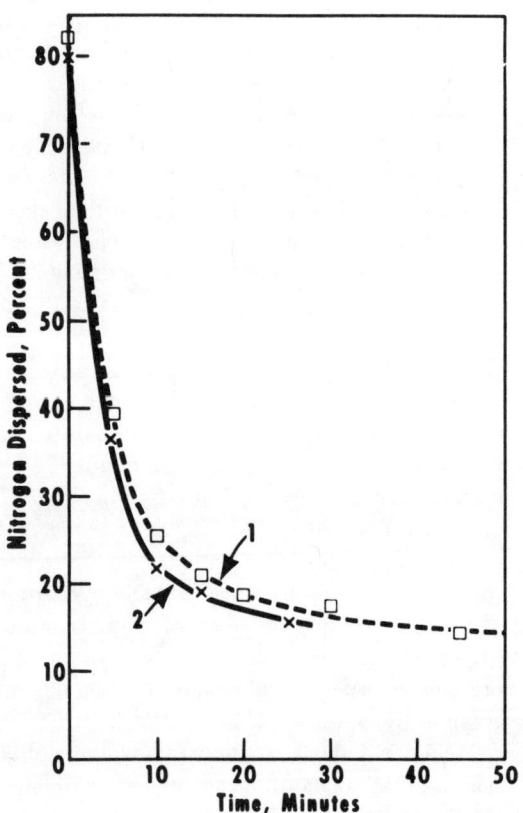

FIGURE 12. Changes in water dispersibility of nitrogeneous constituents of soybean flakes with time of steaming at atmospheric pressure. Curve 1 = defatted flakes and Curve 2 = full-fat flakes. (From Belter and Smith.[115])

TABLE 21

Terminology of Solubility Methods for Assessing Heat Treatment of Soy Flours and Related Products

Term	Abbreviation	Calculation*
% Water soluble nitrogen[†]	WSN	$\dfrac{\text{Ml alkali} \times N \times 0.014 \times 100}{\text{Wt of sample}}$
% Nitrogen solubility index[†]	NSI	$\dfrac{\% \text{WSN} \times 100}{\% \text{ Total nitrogen in sample}}$
% Water soluble protein[†]	WSP	$\% \text{WSN} \times 6.25$
% Protein solubility index[†]	PSI	$\dfrac{\% \text{WSP} \times 100}{\% \text{ Total nitrogen in sample} \times 6.25}$
% Water dispersible protein[‡]	WDP	$\dfrac{\text{Ml alkali} \times N \times 0.014 \times 100 \times 6.25}{\text{Wt of sample}}$
% Protein dispersibility index[‡]	PDI	$\dfrac{\% \text{WDP} \times 100}{\% \text{ Total nitrogen in sample} \times 6.25}$

*Calculations are based on Kjeldahl analysis of extracts where N = normality of alkali, 0.014 = millequivalent weight of nitrogen, 6.25 = nitrogen to protein conversion factor.
[†]Based on AOCS Method Ba 11-65[117]
[‡]Based on AOCS Method Ba 10-65[117]

solubilized on heating beyond the point of minimum solubility presumably is denatured but is altered to a soluble form. Obviously, NSI or PDI are erroneous indicators of extent of denaturation or time of heating under these conditions. An enzymatic hydrolysis method, therefore, was developed to determine extent of denaturation.[120] The method relies on the well-known stability of many native proteins to digestion by proteolytic enzymes and their ease of hydrolysis after denaturation. Denaturation of soybean proteins is important in preparing soy sauce by enzymatic digestion as practiced in Japan since yield depends on completeness of protein hydrolysis. The enzymatic procedure is therefore used to determine whether soybean flakes used for soy sauce are adequately heated.

Heating of commercial soy proteinates at concentrations above 7% causes increases in viscosity and then gelation. Heating for 10 to 30 min at 70 to 100°C is sufficient to effect gelation, but at 125°C the gels are disrupted. Cysteine and sodium sulfite, which act as solubilizing agents for isolates, decrease viscosities of heated and unheated dispersions of the isolates and inhibit gelation.[100] Gels can also be prepared by heating the protein in dilute solution, precipitating with acid, neutralizing, diluting to 20% dispersions, and heating at 90° to 95°C. Sodium bisulfite and mercaptoethanol likewise inhibit gelation under these conditions.[121] Disulfide bonds are therefore an important factor in heat gelation. Intermolecular crosslinks formed by sulfhydryl-disulfide interchange may help stabilize the protein network, or intramolecular disulfide bonds may assist in maintaining conformations of certain molecules, thereby favoring other interactions required for gelation.

A sol-progel-gel transition has been proposed for heat gelation of acid-precipitated globulins.[122] On heating, the sol is irreversibly converted to the progel which in turn gels on cooling. The gel is reversibly converted to the progel by reheating. During conversion from sol to progel, viscosity increases as temperature is raised until a maximum is reached. At higher temperatures the viscosity decreases as a result of irreversible conversion to a metasol state that does not gel on cooling. These changes are summarized as follows:

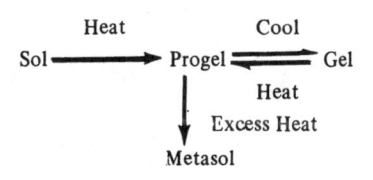

$$\text{Sol} \xrightarrow{\text{Heat}} \text{Progel} \underset{\text{Heat}}{\overset{\text{Cool}}{\rightleftarrows}} \text{Gel}$$
$$\text{Progel} \xrightarrow{\text{Excess Heat}} \text{Metasol}$$

In this scheme the effect of disulfide-cleaving agents is to convert the progel to metasol. Because of the complexity of soybean globulins it is not yet possible to describe the reactions involved in these transformations.

Heating dilute solutions of water-extractable proteins aggregates the 11S and 15S fractions plus part of the 7S fraction as measured by ultra-centrifugation[123] and gel filtration.[124] Solutions of 11S protein heated above 70°C become turbid, and precipitate forms at 90°C. The heated 11S protein is dissociated into subunits as indicated by disc electrophoresis, but heating at 90°C for one hr did not abolish reaction of 11S protein with its antibody.[91]

More detailed studies on effects of heating 11S protein revealed several reactions.[107] Dilute solutions of the protein at 100°C (pH 7.6, 0.5 ionic strength) quickly become turbid and a precipitate forms. Figure 13 shows the changes in solution

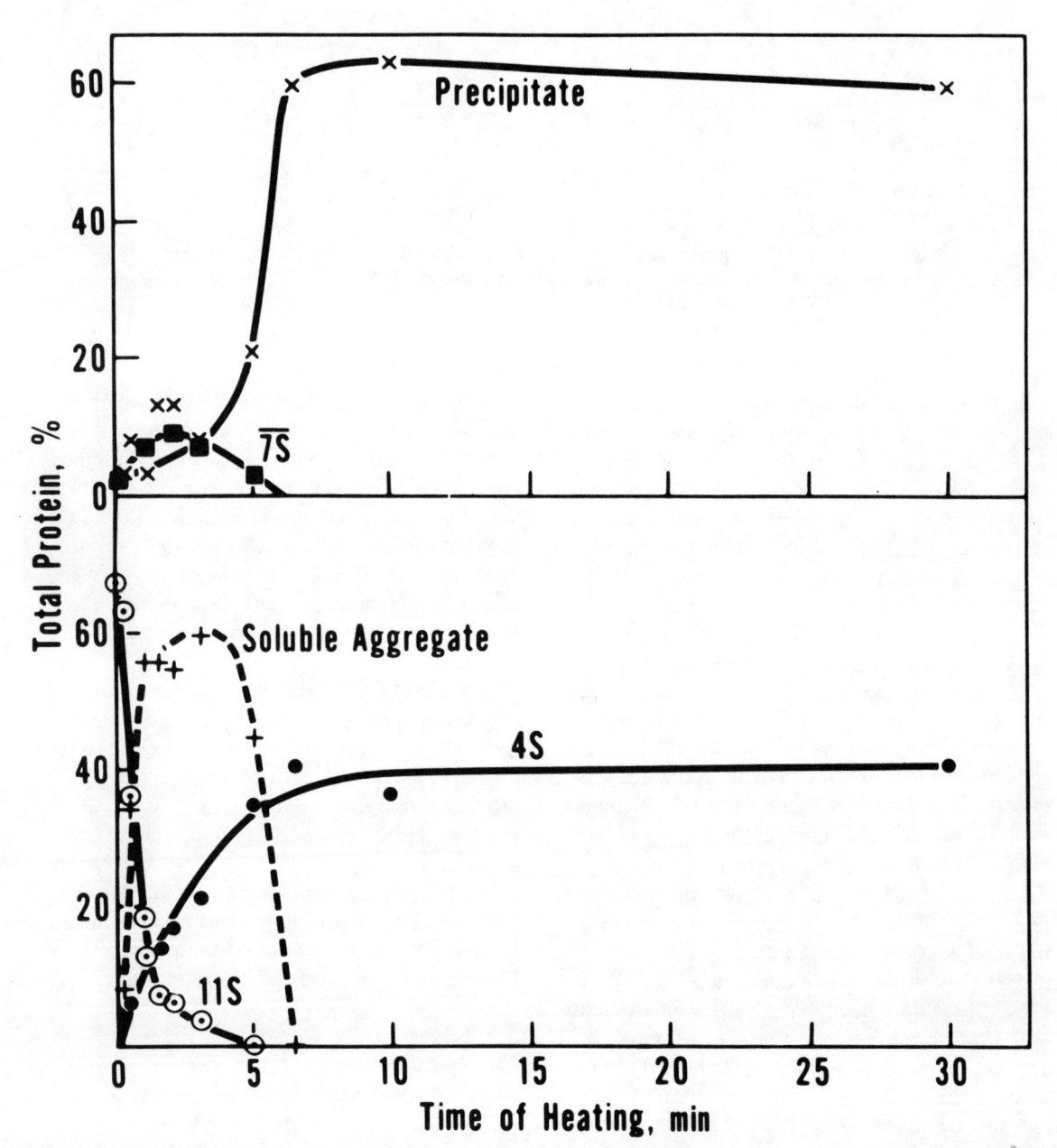

FIGURE 13. Changes in 11S globulin solution as a function of time at 100°C in pH 7.6, 0.5 ionic strength buffer. (From Wolf and Tamura.[107])

32 *Soybeans as a Food Source*

properties as a function of heating time as followed by ultracentrifugation. In less than 5 min the 11S component disappeared completely and a soluble aggregate (80 to 100S) appeared. On further heating the aggregate continued to increase in size until it precipitated. While the 11S component disappeared there was also a slow-sedimenting fraction formed with an $s_{20\,w}$ value of 3 to 4S (labeled 4S in Figure 13). A transient 7S fraction designated $\overline{7S}$ was also detected. The 4S fraction attained a maximum concentration after about 7 min heating and remained constant on heating for up to 30 min. Precipitation occurred more rapidly than indicated in Figure 13 when the solutions were heated with 0.1 to 0.5 M mercaptoethanol and the 80 to 100S soluble aggregate was not detected. When the 11S protein was heated in 0.01 M N-ethylmaleimide, a stable soluble aggregate formed but no precipitate. Disruption of the quaternary structure according to the following scheme is proposed:

$$\text{11S} \xrightarrow{(a)} \text{A-subunits} + \text{[B-subunits]} \xrightarrow{(b)} \text{Soluble aggregates} \xrightarrow{(c)} \text{Insoluble aggregates}$$

In this scheme, A-subunits represent the soluble protein (3 to 4S fraction) and B-subunits represent that part of the 11S molecule which is converted into soluble aggregates by reaction (b). The B-subunits are not detected, presumably because reaction (b) is very rapid as compared to reaction (a). Reaction (c) is the step promoted by mercaptoethanol and inhibited by N-ethylmaleimide. Cleavage of disulfides during reaciton (c) appears to expose groups that undergo hydrophobic bonding to form the precipitate.

Denaturation by Acid or Alkali—Extremes of pH denature soybean globulins. As discussed earlier, high pH disrupts the subunit structures of the 7S, 11S and other globulins, but this process is not reversed when the proteins are adjusted back to neutral pH. When a solution of the globulins which gives an ultracentrifuge pattern similar to Figure 10 is adjusted to pH 12 in 0.5 M sodium chloride, only a 3S fraction is observed. On subsequent adjustment of the pH 12 solution to pH 7.6, 0.5 ionic strength, more than half of the protein precipitates and only a 3S and a 7S peak are found on ultracentrifugation.[110] Exposure of 7S globulin to pH 12 results in irreversible conversion to a 0.4S form.[108] Obviously, protein modification occurs as a result of exposure to pH 12.

When soybean globulins are lowered from pH 3.8 to 2.0 at 0.06 ionic strength, there is a progressive conversion from the four sedimenting fractions detected at pH 7.6 (Figure 10) to 2 to 3S and 7S fractions.[125] These changes are indicative of dissociation of quaternary structures into subunits. Reversibility of these reactions on neutralizing the acid solutions was not tested, but other studies show that irreversible changes occur. Titration of water extracts of defatted meal to pH 2.4 followed by neutralization 2 hr later insolubilized more than a third of the globulins and significantly decreased solubility of all fractions except the 2S fraction. The latter increased, presumably because of 2S-like protein formed by dissociation of other proteins.[112]

The 11S protein is known to break down into 2 to 4S units through a 7S intermediate when 11S solutions are adjusted to low pH and low ionic strength. Although stable at pH 3.8, 1.0 ionic strength, the 11S protein is completely converted to a 2S form at pH 2.2, 0.01 ionic strength.[105] These dissociation reactions of the 11S protein are not reversed by neutralizing the protein solution.[91,102]

In contrast to the sensitivity of the 11S protein, the 7S globulin is more stable to acid. The 7S protein can be dissolved in 0.01 N hydrochloric acid (pH 2) and then brought back to pH 7.6 with retention of its ability to undergo the monomer-dimer reaction with changes in ionic strength. Irreversible changes will occur, however, if the 7S globulin is kept at pH 2 for a long time or if it is dissolved in stronger acid (0.1 N).[108]

Amino Acid Composition

An important quality of soy proteins is their amino acid composition which influences physical, chemical, and nutritional properties. Table 22 lists essential and nonessential amino acids for meal, concentrates, and isolates. Fractionation occurring when meal or flakes are processed into concentrates and isolates accounts for differences in amino acid contents noted between the three different protein forms. Lysine and methionine contents are of major importance. The high lysine content of soy proteins makes them useful for supplementing cereal proteins which are low in

TABLE 22

Amino Acid Composition of Soybean Protein Products

Grams amino acid/16 g

Amino acid	Meal*	Nitrogen concentrate[†]	Isolate[†]
Essential			
Lysine	6.9	6.3	6.1
Methionine	1.6	1.4	1.1
Cystine	1.6	1.6	1.0
Tryptophan	1.3	1.5	1.4
Threonine	4.3	4.2	3.7
Isoleucine	5.1	4.8	4.9
Leucine	7.7	7.8	7.7
Phenylalanine	5.0	5.2	5.4
Valine	5.4	4.9	4.8
Nonessential			
Arginine	8.4	7.5	7.8
Histidine	2.6	2.7	2.5
Tyrosine	3.9	3.9	3.7
Serine	5.6	5.7	5.5
Glutamic acid	21.0	19.8	20.5
Aspartic acid	12.0	12.0	11.9
Glycine	4.5	4.4	4.0
Alanine	4.5	4.4	3.9
Proline	6.3	5.2	5.3
Ammonia	2.1	1.9	2.0

*From Rackis et al.[126]
[†]From technical bulletins.[127,128]

lysine. Methionine is the first limiting amino acid of soy proteins and this deficiency needs to be considered when the proteins are used for nutritional purposes. Cystine content likewise is low, but this amino acid contributes to the gelling properties of isolates as discussed earlier.

Forms of Soy Proteins

Soy protein forms used in foods consist of two groups—whole soybeans and processed soybean protein products.

Whole Soybeans

Whole soybeans are used only in small amounts in U.S. food products such as a soy milk-based infant formula and beverage, canned green soybeans, and canned soybeans in tomato sauce. The infant formula is designed for babies who are allergic to cow's milk while the other products are consumed primarily by vegetarians. In addition, a number of small manufacturers in San Francisco and other Pacific Coast cities convert whole soybeans into Oriental foods.[129]

The largest use of whole soybeans is in the Orient for preparation of foods including soy milk, tofu, miso, and tempeh which are described later. For example, in Japan 642,000 metric tons of soybeans, or 28% of the total soybeans consumed in 1967, were converted into foods based on whole soybeans. For the same year Japan produced only about 4% of its total soybean supply. The remaining soybeans used in Japan were imported from the U.S. and from Communist China.[130] Thus, appreciable amounts of whole soybeans from the U.S. are utilized in foods. Of the soybeans exported to Japan, about two million bushels are selected varieties, including Hawkeye and Kanrich, which are preferred over other varieties for making tofu and miso, respectively.

Processed Soybean Protein Products

The processed soybean protein products used as food ingredients in the U.S. are conveniently divided into three categories based on protein content: soy flours and grits, protein concentrates, and protein isolates. These three types of soy proteins supplied by the soy industry are considered the starting materials for the variety of soy protein products now available. In certain instances these materials may be processed further before they are incorporated into a food product, e.g., extrusion of soy flour or spinning of isolates to form simulated meats. These additionally processed products are considered in a later section (Functional Properties).

Soy Flours and Grits—Most soy flours and grits are prepared from defatted flakes and have similar chemical compositions. All edible grits and flours are made from dehulled beans; a major difference among them is particle size. The following classification is used to differentiate these products:

Product	Mesh size*
Grits	
Coarse	10–20
Medium	20–40
Fine	40–80
Flours	100 or finer

*U.S. standard screens

Grits are obtained by coarse grinding and screening while flours are prepared by fine grinding until most of the product passes through a 100-mesh screen. Most soy flours, however, are ground to 200 mesh and certain specialty flours are ground until

TABLE 23

Proximate Analyses of Flours and Grits*

	Defatted	Low-fat	Full-fat
Protein (N × 6.25), %	51	46	41
Fat, %	1.5	6.5	21
Fiber, %	3.2	3.0	2.8
Ash, %	5.8	5.5	5.3
Carbohydrates, %	34	34	25

*As is basis with normal moistures 5 to 10%.
(From Horan.[131])

most of the flour passes through a 300-mesh screen. The term flour as applied to soy thus refers only to particle size; soy flour proteins do not have the viscoelastic properties characteristic of wheat flour proteins.

Flours and grits vary in fat content and extent of heat treatment as well as particle size. This variety of products has been developed to provide the array of physical and functional properties needed for incorporation of soy proteins into many different foods. Flours and grits are the least refined of soy proteins and hence the lowest in protein content. Typical analyses are given in Table 23.

Defatted flours and grits are prepared by hexane extraction of flakes as outlined earlier (Figure 5). The treatment the flakes receive after they leave the extractor determines the properties of the final products and their ultimate use. Flakes coming from the extractor contain about 30% hexane, and removal of solvent is concerned with controlling three variables that influence protein denaturation: time, temperature, and moisture content.

Three main types of equipment are available[132] for removing hexane from extracted flakes for the preparation of edible grade flakes:

(a) Schneckens system
(b) Flash desolventizer
(c) Vapor desolventizer-deodorizer

The Schneckens system (Figure 14) consists of a series of steam-jacketed horizontal conveyors arranged one above the other in banks of about five units. Wet flakes enter at the top and proceed downward through the heated units, thereby vaporizing the hexane. Steam is introduced into the bottom unit to strip off residual solvent. A disadvantage of this processing is that the desolventized flakes will have a maximum PDI of only 65 to 70 whereas normal PDI values are in the range of 40 to 50. Other drawbacks of this equipment are difficulties of cleaning, high maintenance costs, and inefficiency of solvent removal.

Flash desolventization is a newer process than the Schneckens system and has the advantage of producing flakes of high protein solubility.[133] A combination flash desolventizer-deodorizer is outlined in Figure 15. Wet flakes enter a rapidly moving stream of superheated hexane which flash evaporates the hexane from the flakes and also carries them to a cyclone separator. Hexane vapors leave the top of the cyclone and are split into two streams. One stream is condensed and removed while the other is recycled through the superheater and desolventizer tube. Desolventized flakes leave the bottom of the cyclone through a rotary valve and enter the deodorizer where residual solvent is removed by upward flow of an inert purge gas such as carbon dioxide or nitrogen. After passing thorugh a condenser to remove the hexane, the purge gas can be recycled through the deodorizer. Flakes from this operation will have PDI values in the range of 70 to 90 depending on operating conditions, but the equipment can be operated so that PDI values of desolventized flakes will be only 1 to 3 units below the value for the flakes as they leave the extractor.

In the vapor desolventizer-deodorizer (Figure 16), the wet flakes fall through a stream of superheated hexane to remove the bulk of the solvent as in the flash desolventizer. The flakes then move to the opposite end of the desolventizer in 3 to 4 min where they leave through a rotary valve. A loss of only 1 to 2 PDI units occurs in the desolventizer. The flakes can then be passed into the deodorizer where residual solvent is stripped off by passing steam through them while they are conveyed through the unit. If the deodorizer is operated under vacuum (1/2 atmosphere) the steam does not condense on the flakes but merely strips out the hexane and yields flakes with PDI values only 1 to 2 units lower than the starting flakes. Such flakes are ideally suited for preparation of isolates since yields will be higher than with flakes of lower PDI values. Alternatively, if flakes of lower PDI values are desired for soy flours and grits, the deodorizer is operated at 1 to 2 atmospheres pressure; under these conditions some of the steam condenses on the flakes. The combination of heat and moisture now causes denaturation of the proteins with accompanying

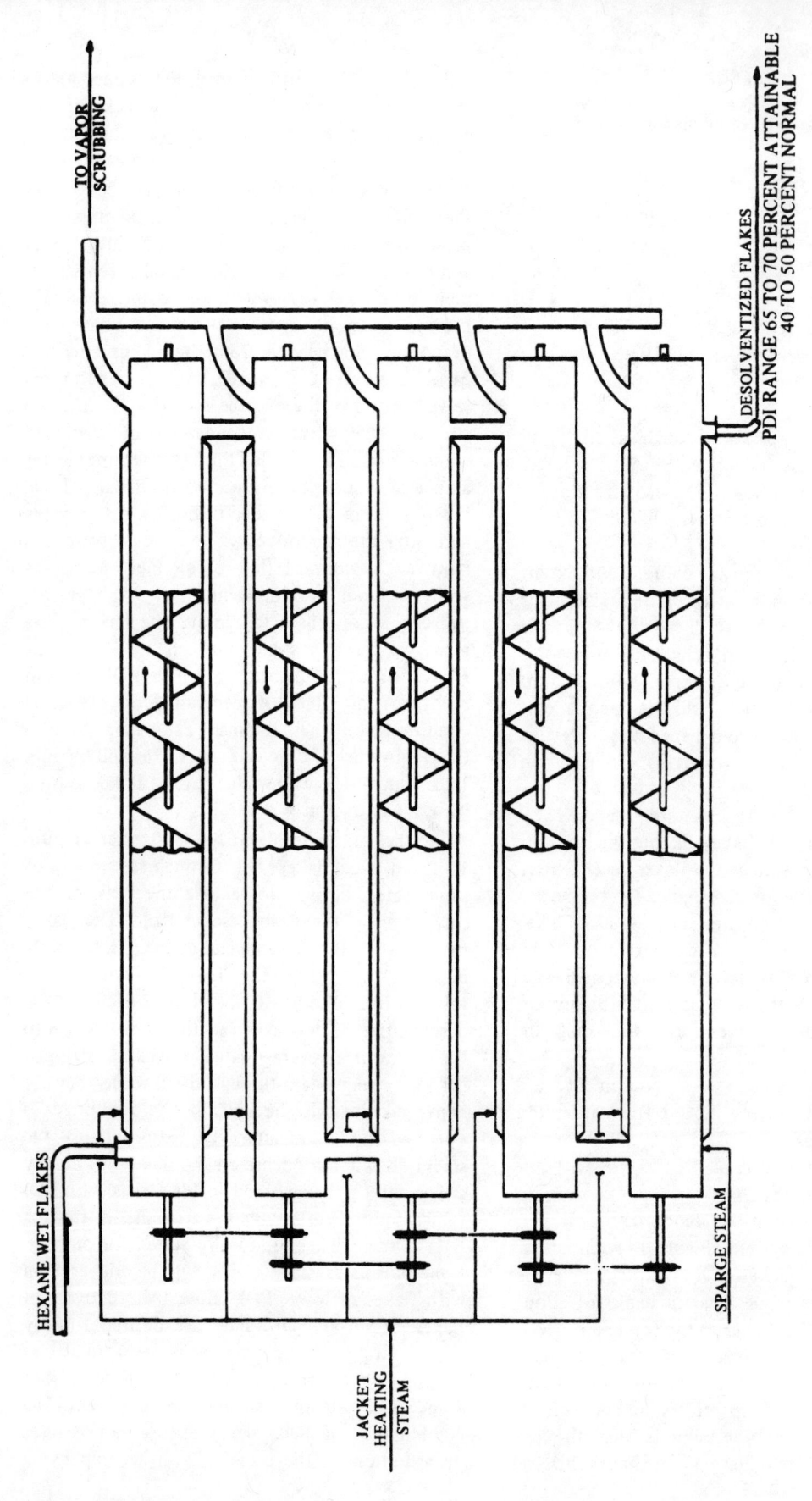

FIGURE 14. Diagram of Schneckens system for desolventizing soybean flakes. (From Blaw-Knox Chemical Plants, Inc. With permission.)

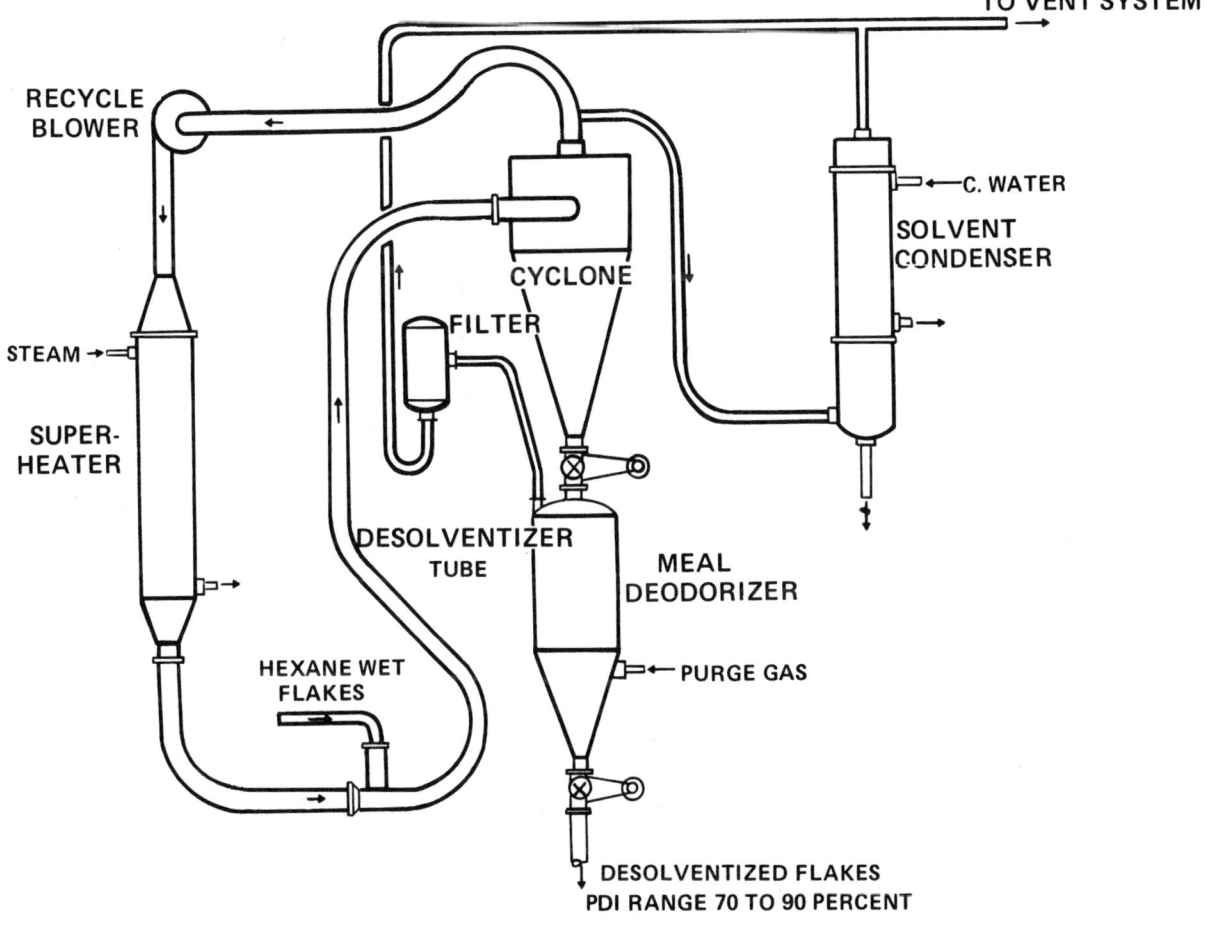

FIGURE 15. Diagram of flash desolventizer-deodorizer system for removing hexane from soybean flakes. (From Blaw-Knox Chemical Plants, Inc. With permission.)

changes in functional properties and improvements in nutritive value and flavor. If additional moisture is desired, it can be added by spraying water on the flakes as they pass through the mixer-conveyer between the desolventizer and deodorizer. By increasing moisture content and residence time in the deodorizer, completely cooked flakes (PDI of 10) are produced. This type of operation is therefore very flexible and permits production of defatted flakes with PDI values ranging from 10 to 90.

Low-fat products are prepared by one of three ways: (a) screw pressing of cracked full-fat meats; (b) mixing of full-fat flours with defatted flour; and (c) addition of oil to defatted flour. Also available are high-fat and lecithinated soy flours. High-fat flours are prepared by adding soybean oil to defatted flour in the range of 15%. Lecithinated flours are low-fat or high-fat flours with added lecithin, usually at a level of 15%.

Full-fat soy flours are prepared by steaming soybeans to eliminate the beany and bitter flavors characteristic of raw soybeans and to inactivate lipoxygenase and other enzymes before crushing the seed. The beans are then dried to below 5% moisture, passed through cracking rolls, dehulled, and finally ground to 200 mesh. This product has a PDI of 35 to 45.

The Northern Regional Research Laboratory has recently developed a process for preparing full-fat flours by extrusion cooking (Figure 17).[134] Cracked, dehulled beans are heated dry to inactivate lipoxygenase, tempered to a desired moisture level, and then passed through the extruder. After cooling and grinding, a full-fat flour of good nutritional quality is obtained. The

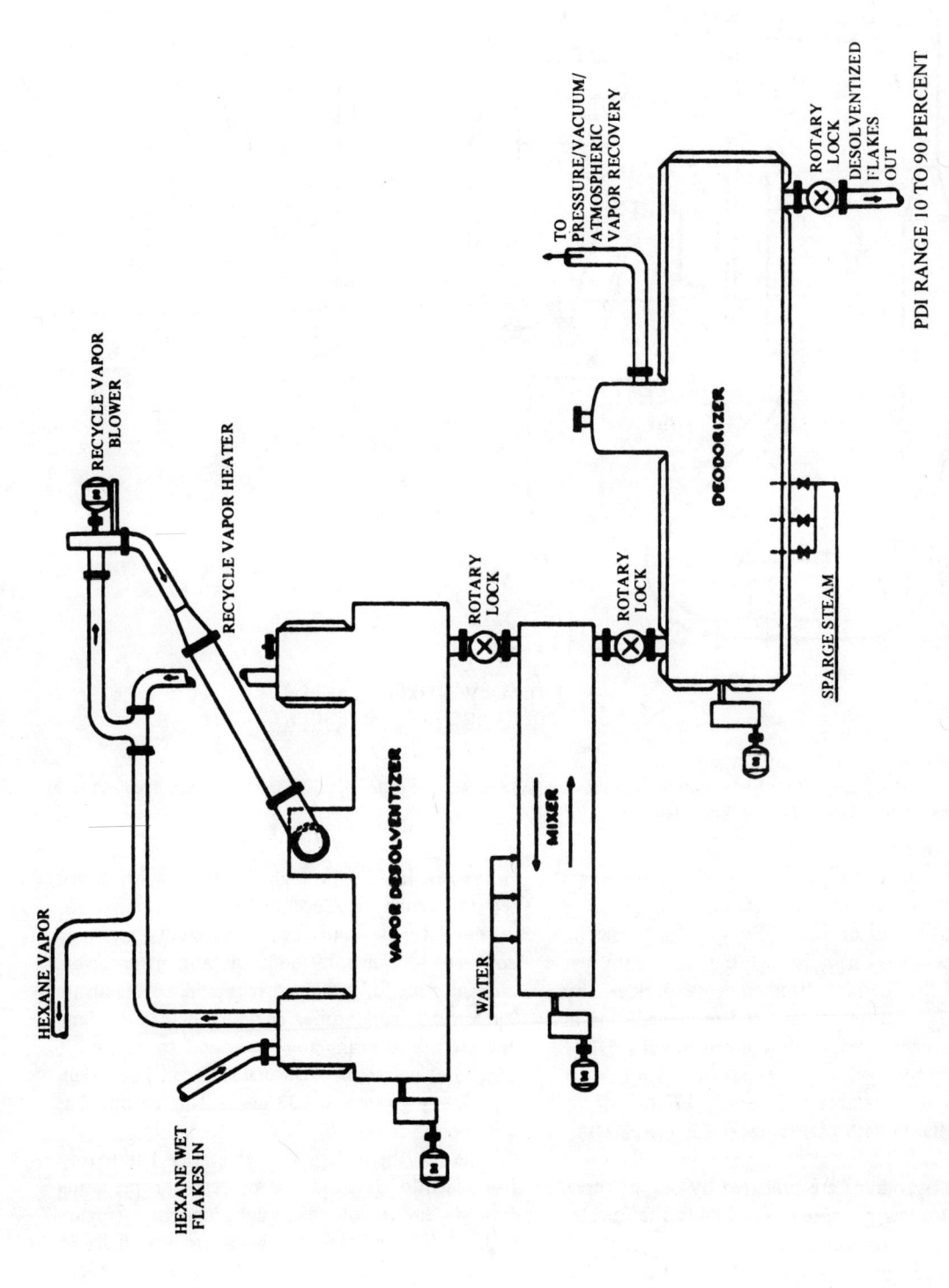

FIGURE 16. Vapor desolventizer-deodorizer system for removing hexane from soybean flakes. (From Blaw-Knox Chemical Plants, Inc. With permission.)

process is essentially a continuous pressure cooking operation. The rotating screw forces the soybeans through a small die constriction causing a rise in pressure and temperature which results in rapid cooking with attendant inactivation of antinutritional factors and elimination of the beany and bitter flavors of raw soybeans.

Protein Concentrates—Flours and grits have protein contents in the range of 40 to 50% (Table 23). To prepare products of higher protein content it is necessary to further process flakes and flours to remove some of the low-molecular-weight components. Protein concentrates with minimum protein contents of 70% are obtained by removing the water-soluble sugars, ash, and other minor constituents of defatted soy flakes or flours. Three processes are used commercially to prepare concentrates. The processes differ mainly in the methods used to insolubilize the major proteins while the low-molecular-weight components are removed (Figure 18). In the first process the nonprotein constituents are extracted with aqueous alcohol, leaving the proteins and polysaccharides which are desolventized and dried to yield the concentrate.[135,136] In the second process the major proteins are insolubilized by extracting with dilute acid at pH 4.5 (the isoelectric point of the proteins, Figure 9). Since some of the minor proteins are soluble at pH 4.5, there is some loss of protein in this process. After the acid leach, the insoluble polysaccharide-protein mixture is adjusted to near neutrality and dried.[137,138] The third process takes advantage of the heat-sensitivity of soy proteins; the flakes or flour are heated with moisture to denature the proteins and insolubilize them. The low-molecular-weight constituents are then extracted with water.[139]

Compositions of the three types of concentrates are shown in Table 24. Despite the different processes used, overall compositions of the concentrates are very similar; protein content on a dry basis varies only from 71 to 72%. The major nonprotein constituents in concentrates are polysaccharides-arabinogalactan, acidic pectin-like polysaccharide, arabinan, and cellulose.[141] Cellulose is believed to arise from residual soybean hull. One obvious difference between the three concentrates is in water solubility of the proteins. The proteins in the alcohol leached and moist-heat water leached concentrates are denatured and insoluble, while protein insolubilization is much less in the acid leached concentrate. The higher protein solubility of the acid leached concentrate is claimed to be advantageous in certain food applications.

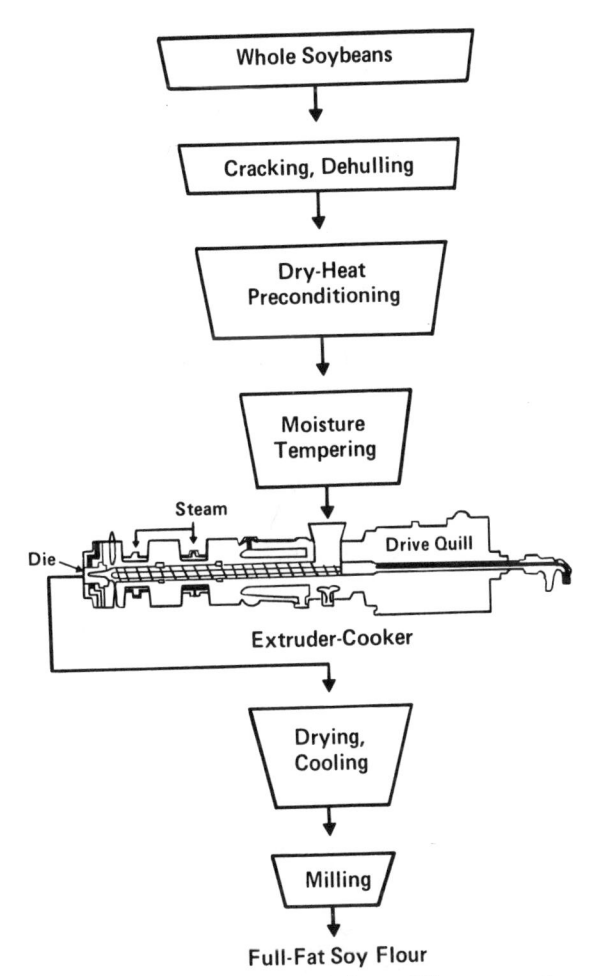

FIGURE 17. Outline of extrusion cooking process for preparing full-fat soy flour. (From Mustakas et al.[134])

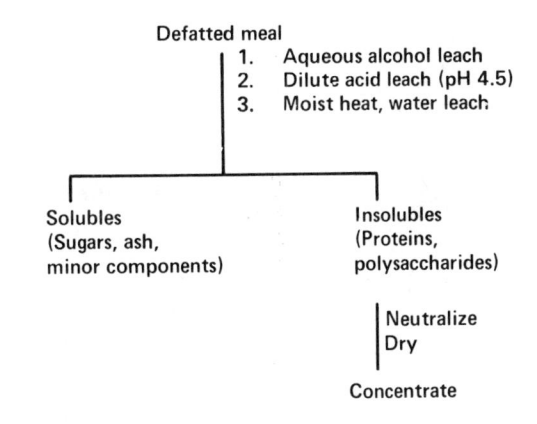

FIGURE 18. Outline of processes for preparing soy protein concentrates. (From Wolf.[77])

Isolates—Most refined forms of soy proteins are the isolates. They are processed one step further than the concentrates by removing the water-insoluble polysaccharides as well as the water-soluble sugars and other minor constituents (Figure 19). Three U.S. companies manufacture and market isolates, but the processes are similar in principle. Defatted flakes or flour of high protein solubility are extracted with dilute alkali (pH 7 to 9) at 50 to 55°C. The extract is then separated from the insoluble residue (water-insoluble polysaccharides plus residual protein) by screening, filtering, and centrifuging. On adjusting the pH of the extract to about 4.5 (the isoelectric region, Figure 9), with food-grade acid the major proteins precipitate. After filtering and/or centrifuging the protein curd from the solubles (whey), it is washed with water. Direct spray drying of the curd yields the isoelectric protein while neutralization followed by spray drying gives the proteinate form of the isolate. Proteinates are usually preferred because they are water-dispersible and therefore easier to incorporate into food products. Moreover, solubility is usually necessary to obtain the functional effects of the proteins. Sodium proteinates are the major form of isolate sold, but potassium and calcium proteinates are also available.

As a rule of thumb each of the three fractions obtained in the isolate process—residue, whey

TABLE 24

Proximate Compositions of Soy Protein Concentrates

	Alcohol leach	Acid leach	Moist heat, water leach
Protein (N × 6.25), %	66	67	70
Moisture, %	6.7	5.2	3.1
Fat (Petr. ether extractable), %	0.3	0.3	1.2
Crude fiber, %	3.5	3.4	4.4
Ash, %	5.6	4.8	3.7
Nitrogen solubility index	5	69	3
pH of 1:10 water dispersion	6.9	6.6	6.9

(From Meyer.[140])

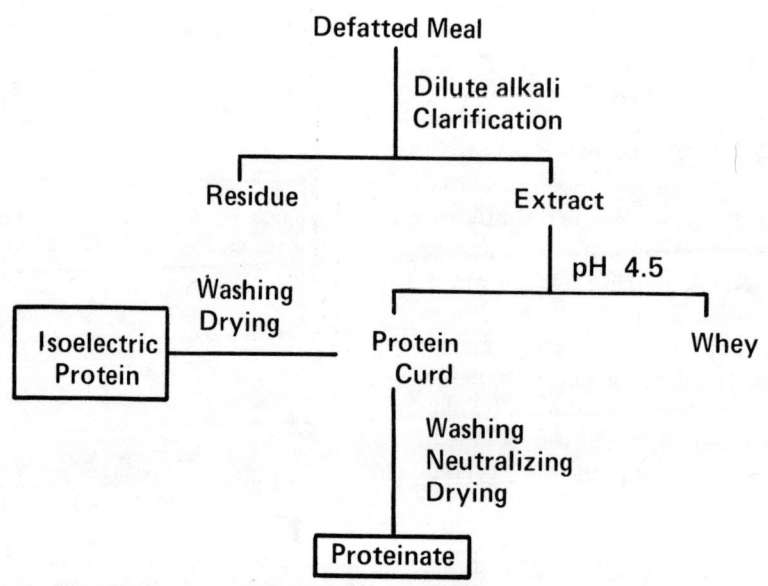

FIGURE 19. Diagram for commercial production of soybean protein isolates. (From Wolf.[77])

TABLE 25

Proximate Analysis of Commercial Soy Protein Isolates

	A	B	C	D
Protein, %	92.8	92.2	92.9	94.7
Moisture, %	4.7	6.4	7.6	3.7
Crude fiber, %	0.2	0.1	0.1	0.2
Ash, %	3.8	3.5	2.0	2.7
Nitrogen solubility index	85	95	—	—
pH (1:10 aq. dispersion)	7.1	6.8	5.2	5.5

(From Meyer.[140])

solids, and isolate—represents roughly one third of the starting flakes when isolation is carried out on a laboratory scale. In commercial practice, however, conditions giving maximum yield of isolate are not necessarily optimum from an economic standpoint. Patent literature indicates that yields of 30 to 32% of the weight of the starting flakes are obtainable,[142] but on a large commercial scale yields of isolate are probably lower.

Proximate analyses of four commercial soy protein isolates are shown in Table 25. Samples A and B are of the proteinate type while C and D are in the isoelectric form. The latter have a lower pH, lower ash, and are insoluble in water. Protein contents on a dry basis range from 97 to 100% calculated with a nitrogen-to-protein conversion factor of 6.25 (assuming a nitrogen content of 16%) which is commonly used in industry. There is no question that this factor is too high since nitrogen contents of purified acid-precipitated proteins are higher than 16% nitrogen,[143,144] but there is no agreement on what the correct factor is. Laboratory and pilot-plant preparations of isolates contain 0.5 to 1.1% phosphorus.[113] Thorough dialysis of isolates reduces the phosphorus level to about 0.2%[145] (residual phosphorus probably consists of ribonucleic acids and phosphatides); the dialyzable phosphorus is believed to be phytate. A typical commercial isolate contains 0.8% phosphorus.[128]

Laboratory preparations of isolates also contain alcohol-extractable materials consisting of glycerides, phosphatides, saponins, sterol glycosides, and isoflavones. A yield of 3.6% of alcohol-extractables was obtained from a dialyzed isolate.[145] It is likely, however, that more non-protein materials are present because additional saponins were isolated from alcohol-extracted protein.[146]

These non-protein materials affect physical and chemical properties of isolates. For example, alcohol-washed isolates form very stable foams while the unwashed isolates do not.[144,147] The alcohol-extractable materials were also noted to darken on exposure to air and to develop rancid odors, presumably because of autoxidation of the lipids.[145] It has not been demonstrated that the residual lipids and other non-protein constituents contribute to the flavor of isolates, but they are likely suspects for flavor defects arising when isolates are incorporated into certain food products. Painty flavors, for example, have been reported for spun fibers prepared from isolates.[148]

Although chemical compositions of isolates are similar, physical properties such as solubility and molecular-weight distribution may differ significantly. As discussed previously (Solubilities of Isolates), studies on laboratory and commercial preparation of isolates revealed large variations in solubility in pH 7.6 phosphate-sodium chloride buffer.[99] One isolate had a high solubility but did not give an ultracentrifugal pattern characteristic of laboratory preparations; one third of the protein was in a low-molecular-weight form. Heat treatment given the protein during one of the isolation steps is believed responsible for the changed distribution of molecular weights. Proximate analyses are therefore not sufficient for selection of commercial isolates for a given application.

Essential amino acid content of a commercial isolate is given in Table 22. Isolates tend to be lower in lysine and the first limiting amino acids—methionine and cystine—than concentrates or soy flours.

Selling Prices and Production Estimates

The three forms of soy protein are sold at prices that reflect the extent of processing received by the products (Table 26). Flours and grits are lowest in price and on a price-per-pound-of-protein basis are a very economical source of protein. Concentrates sell for two to three times the price of flours while isolates are priced about fivefold higher than flours. Prices of soy proteins in their various forms have held relatively stable despite inflationary trends in recent years. In 1966 it was predicted that over the next five years, with increased volume of production and further improvements in technology, concentrates selling for 18 to 28 cents could drop to 14 cents per pound while isolates selling at 35 to 38 cents could

TABLE 26

Selling Prices and Production Estimates for Soybean Proteins

Protein form	Protein content, %	Price per pound, cents	Price per pound of protein, cents	Annual* production, million pounds
Flours and grits (defatted)	50	7–8	14–16	232–237
Concentrates	70	18–26	26–37	20–35
Isolates	97	35–45	36–46	25–40

*Estimate for 1970.

(From revised data of Eley[149] and Wilding.[150])

decrease to 28 cents per pound.[140] Failure of the predicted price decreases to occur can be attributed to increased costs resulting from inflation and a lack of the necessary volume of production.

Precise production figures for the various soybean proteins are unavailable since manufacturers of these products have made large investments in developing processes and markets and, as a result, they are reluctant to reveal exact figures. Estimating the amounts of soy flours and grits used in foods is further complicated because edible grades of flours and grits are purchased from the soy industry and used by others to manufacture non-food items such as pet foods and calf starters, as well as food items.

Production estimates shown in Table 26 are based on data compiled for 1967 and updated to 1970 on the basis of growth rates for 1965 to 1968. Available for over 30 years, soy flours and grits are produced in the largest amounts and find uses in a variety of foods (Table 27). The largest single domestic application is in baked goods followed by meat products and beverages.

Rate of increase in food uses of soybean protein was estimated to by 5 to 7% annually in 1965 to 1967 and was expected to continue at this rate until 1970. Increasing amounts of soy proteins are likely to be used for protein supplementation of cereal and other foods. Thus, a greater rate of growth can be expected for the next few years. Evidence for expanded use of soy products is the announcement by a producer of soy protein concentrates that their production capacity will be more than doubled by late 1971.

A new use of full-fat soy flour consists of blending it with cheese whey and spray-drying the mixture to produce a powder of good nutritive value suitable for a variety of food uses.[151]

Although details of the process were published only in 1969, several companies marketed whey-soy flour blends in 1970. Because large quantities of cheese whey are available and because of more stringent regulations against pollution of streams with whey, this use of soy flour and other soy proteins may develop into a sizable market.

One use of soy protein that is expected to grow rapidly is in simulated meat products prepared by the extrusion or spun fiber technique. In 1969 annual sales were estimated at ten million dollars but were predicted to rise to two billion dollars by 1980.[152] One company completed a multimillion-dollar plant in 1970 for preparation of spun-fiber protein foods[153] and at least five concerns are manufacturing extruded soy flour products.

TABLE 27

Uses of Soy Flours and Grits

Use category	Million pounds*
Domestic	
Baked goods	60
Meat products	38
Beverages	12
Dry cereals and infant foods	8
Brewers' flakes	4
Pasta and macaroni products	1
Miscellaneous	7–12
Subtotal	130–135
Exports	12
U.S. Government†	90
Total	232–237

*Estimate for 1970.
†Purchase of corn-soy-milk (CSM) blend for overseas distribution.

(From revised data of Eley[149] and Wilding.[150])

Functional Properties

Until the recent development of textured soybean proteins, the major reason for adding soy proteins to foods in the U.S. was for their functional properties rather than as a source of dietary protein. Nonetheless, the functional properties and the cost advantages of soy proteins were a sufficient basis for establishing a definite pattern of usage in a variety of processed foods. Several recent reviews on functional properties of soybean proteins are available.[77,154-156]

A large number of functional properties are attributed to soy protein but a critical review of these properties is difficult because supporting data are often meager or lacking altogether. In some instances the functional properties claimed for a given soy protein product are known only to the food processor using it.

There are no standard tests available for measuring functional properties of soy proteins. A few methods have been proposed, but they are generally empirical in approach and may have limited usefulness. For example, two methods have been proposed for measuring emulsification properties of soy proteins as meat additives, but only one of the tests is claimed to give results in agreement with manufacturing practice.[157] Crude correlations have been found between functional properties and protein solubility tests. Thus, NSI or PDI determinations are often used as quality control tests in preparing products such as soy flours for certain functional uses. It is likely, however, that if differing processing conditions were used to arrive at the same NSI or PDI value for two different soy flours, their functional properties would not agree. It should be stressed that the only conclusive test for a given functional property is the incorporation of the soy protein product into a finished food.

Functional properties of soy proteins are listed in Table 28. Although these functional properties are usually attributed to the proteins, in the cruder forms other components may also influence the effects observed. In flours, grits, and concentrates, for example, the polysaccharides as well as the proteins will absorb water; consequently, these products absorb more water than an equivalent amount of protein in the form of an isolate. In the preparation of simulated meats by extrusion of defatted soy flour, the carbohydrates allow the product to expand or puff to give a fibrous, lacy network that is desirable for proper texture. If the extruded product contains more than about 75% protein, it will have a gummy texture.[158]

Emulsification

Soy proteins probably play two roles in emulsification. They may aid in the formation of oil-in-water emulsions and stabilize the emulsions once they are formed. Since proteins are surface active, they collect at oil-water interfaces and lower surface tension, thereby making it easier to form emulsions. The emulsified oil droplets are then stabilized by collection of proteins at the surface of the droplets to form a protective barrier that prevents their coalescence and emulsion breakdown. Emulsion stability is important since success of an emulsifier depends on its ability to maintain the emulsion in subsequent processing steps such as cooking and canning.

Soy flours, concentrates, and isolates are used extensively as emulsifiers in ground meat products. Emulsification tests in a model system of soybean oil and water suggested that protein isolates would be poor emulsifiers in ground meat products.[164] However, in actual sausage manufacture two soy sodium protein isolates were superior to sodium caseinate while a soy protein concentrate (produced by the alcohol leach process) with a low protein solubility was a poor emulsifier for this purpose (Table 29). Emulsification by protein isolates also depended on their NSI values; an isolate with an NSI value of 32 did not give stable emulsions in luncheon meat while isolates with NSI values of ~ 80 were satisfactory. Based on these practical results an emulsion stability test was devised and found to correlate with manufacturing experience (Table 29). The two isolates that performed satisfactorily in sausages gave oil separations of about 42% while caseinate and protein concentrate gave higher oil separations and were rated poor to ineffective as stabilizers.

Soy flours are also used as emulsifiers in baked goods and soups. In creamed soups, for example, full-fat soy flour supplies additional fat in a fine dispersion that remains stable during canning, freezing, storage, and on subsequent reheating when it is served.[161]

Fat Absorption

Soy proteins are used for two different purposes in regard to fat absorption in foods. In ground meats soy proteins promote fat absorption or fat-binding, thereby decreasing cooking loss and

TABLE 28
Functional Properties of Soybean Proteins

Property	Protein form used*	Food system	References
Emulsification			
Emulsion formation	F,G,C,I	Frankfurters, bologna, sausages	156, 157, 159, 160
	F	Breads, cakes, soups	161, 162
	I	Whipped toppings, frozen desserts	163
Emulsion stabilization	F,G,C,I	Frankfurters, bologna, sausages	156, 157, 159, 164
	F	Soups	161
Fat absorption			
Promotion	F,G,C,I	Frankfurters, bologna, sausages, meat patties, simulated meats	156, 159–161, 165
Control	F,I	Doughnuts, pancakes	149, 155, 166
Water absorption			
Promotion	F,C	Breads, cakes, confections, simulated meats	161, 162, 165, 167
Control	F	Macaroni	168
Retention	F,C	Breads, cakes, confections	161, 162, 166, 167
	C	Meat patties	160
Texture			
Viscosity	F,C,I	Soups, gravies, chili	161, 166
Gelation	F,C,I	Gound meats	156, 159
	I	Simulated ground meats	169
Shred formation	F,I	Simulated meats	166, 170
Chip and chunk formation	F	Simulated meats, fruits, nuts, and vegetables	158, 165, 166, 171
Fiber formation	I	Simulated meats	171, 172
Spongy structure formation	I	Simulated meats, dried tofu	130, 173, 174
Dough formation	F,C,I	Baked goods	163
Adhesion	C,I	Sausages, luncheon meats, meat patties, meat loaves and rolls, boned hams	156, 159, 166
Cohesion	F,I	Baked goods	163
	F	Macaroni	168
	I	Simulated meats	170
	I	Dried tofu	130, 174
Elasticity	I	Baked goods	163
	I	Simulated meats	170
	I	Gels	100
Film formation	I	Frankfurters, bologna	159, 163, 166
Color control			
Bleaching	F	Breads	161
Browning	F	Breads, pancakes, waffles	149, 161
Aeration	I	Whipped toppings, chiffon mixes, confections	147, 163, 166

*F,G,C, and I represent flours, grits, concentrates, and isolates, respectively.

helping to maintain dimensional stability in the cooked product. One supplier of frozen, pre-cooked beef patties adds protein concentrate during grinding and thus reduces loss of fat and juices during broiling or frying.

Data on fat absorption by soy protein products are meager. Textured soy flours have oil absorptions ranging from 65 to 130% of their dry weight.[165] Maximum absorption occurs in 15 to 20 min and varies with particle size; small particles absorb more oil than large particles.

The mechanism of fat absorption or "binding" has not been explained. In a ground meat product such as frankfurters or luncheon meat, fat "binding" by soy protein appears to involve formation and stabilization of an emulsion plus the formation of a gel matrix which hinders migration of fat to the surface.[159] Fat absorption may thus merely be another aspect of emulsification.

In other foods such as pancakes and doughnuts, the addition of soy flour helps prevent excessive fat absorption during frying.[155] Data for the effect of soy flour on fat absorption in doughnuts are shown in Table 30. Although fat absorption was reduced 60% by the soy flour of high NSI, such flours impart a beany flavor to the doughnuts and, therefore, are not used by bakers. Instead, a compromise is made between functionality and flavor of the doughnuts, and flours with NSI values of 50 to 65 are used. The protective effect of soy proteins in controlling fat absorption during frying has not been explained. The effect may involve heat denaturation of the proteins to form a fat-resistant barrier at the doughnut surface since a soy flour of high NSI performs better than a flour of low NSI in which the proteins are already denatured.

Water Absorption

Soy proteins contain numerous polar side chains along their peptide backbones, thereby making the proteins hydrophilic; consequently, the proteins absorb water and tend to retain it in finished food products. Some of the polar sites such as carboxyl and amino groups are ionizable; thus, polarity can be changed by varying pH. Changing pH alters the water absorption properties of soy flour. For example, at pH 8.5 a dough-like mass of soy flour absorbs about twice as much water as a dough in the pH range of 4.5 to 6.3.[139] Water retention of soy protein gels (measured by determining water loss on compressing gels be-

TABLE 29

Emulsion Stability of Protein Additivies for Meat Products

Phase separation, %

Protein additive	NSI*	Fat in sausages	Oil in emulsification test
None	–	8.2	–
Soy protein isolate A	85.2	0.3	42.4
Soy protein isolate B	82.7	0.4	42.7
Sodium caseinate	97.8	2.3	54.5
Soy protein concentrates	10.8	7.9	96.8

*NSI = nitrogen solubility index.
(From Inklaar and Fortuin.[157])

TABLE 30

Control of Fat Absorption in Doughnuts with Soy Flour

	NSI of soy flour	Fat absorption during frying, g fat/100 g dry mix
Wheat flour mix	–	27.6
Wheat flour mix plus 4% soy flour*	60	22.8
Wheat flour mix plus 4% soy flour*	80	11.1

*Based on weight of wheat flour.
(From Johnson.[155])

tween glass plates) as a function of pH resembles the pH-solubility curve of soy protein (Figure 9); water retention is at a minimum at pH 4.5 and increases rapidly as pH is increased or decreased from this region.[175]

The water absorption properties of soy flours are important in baked goods. If soy flours are substituted for nonfat dry milk solids, it is necessary to add more water to the doughs; this increases dough yield and improves dough handling characteristics.[167] In cakes, soy flours help to minimize shrinkage.[162] When isolates are used in bread, water absorption increases 1 to 1.5% per 1% addition of isolate; calcium-precipitated isolate has a higher water absorption than acid-precipitated protein.[176]

Rate of water absorption of soy grit products depends on particle size and distribution. In some uses it may be desirable to have a final product of high viscosity, but during processing a low viscosity is advantageous to make the product easier to handle. Preparation of canned pet foods has been cited as an example where it is preferred to have a slurry during high-speed canning to make it easy to fill the cans, but absorption of the water must occur after the canning operation to develop the desired consistency.[155] Specially processed soy grits with slow water-absorption characteristics are used for this purpose. In such applications fine particles are undesirable because they absorb water too rapidly and lead to a mushy texture rather than the more desired chunky product.

Water absorption is also important in developing the textured properties of simulated meats prepared by extrusion of soy flour. Water absorption of these materials varies from 2.4 to 3.4 times their dry weight. Hydration is necessary to convert the crunchy texture of the dry products to a fibrous chewy state resembling meat. Rate and extent of water absorption vary with particle size.[165]

Soy protein concentrates prepared by alcohol leaching of defatted flakes have water absorptions of 3.4 to 3.8 times their dry weight.[136] A recent report describes the water-binding properties of soy protein concentrates when added to ground meats.[160] Addition of 8 lb of concentrate, 20 lb of water, and 1 lb of flavoring to 100 lb of hamburger yielded 98.5 lb of cooked product (meat patty) as compared to only 70.2 when hamburger without additives was cooked. The cooked meat patty contained 20.7% protein and 17.9% fat while the cooked hamburger contained 22.8% protein and 20.6% fat. Calculations showed that a 2.75 oz cooked patty could be prepared for the same cost as a 2 oz cooked hamburger because of less cooking shrinkage with the meat patty.

When added to macaroni, a chemically treated soy flour decreases water absorption of the product during cooking. The cooked product is therefore firmer and more resistant to overcooking. This is desirable for canned products and for institutional feeding.[168]

Addition of soy proteins such as flours to baked goods and confections increases water absorption but also helps retain moisture in the products, thereby keeping them fresh longer. This is important in items of short shelf life such as bread and cakes.[177]

Information on tests for water absorption is not readily available. Empirical tests have been developed by different companies in the industry, but these tests are not generally revealed to others. One type of test involves stirring a given weight of soy flour in a known volume of water, centrifuging, and measuring the volume of supernatant. The difference between the amount of added water and the recovered water is used as a measure of absorbed water. Procedures for measuring water retention in ground meats mixed with protein concentrates before and after cooking were described recently.[160] For the uncooked product, a weighed portion is placed in a Petri dish between layers of filter paper, placed in a refrigerator overnight, and reweighed. Weight or "drip" loss is expressed as percent of original weight. Water losses during cooking are determined by sealing a weighed portion of ground meat in a plastic pouch, immersing the pouch in a boiling water bath for 4 min. After cooling, the liquid is withdrawn with a syringe, transferred to a graduated cylinder, and volumes of fat and water are then measured.

Texture

The ability to provide texture to a variety of traditional foods as well as new foods now being developed is a most important property of soy proteins. Texture is provided by soy proteins in a number of ways; one of the simplest is merely to thicken products such as soups and gravies by adding soy flours.

Gelling properties of soy proteins contribute to texture in ground meat products such as frankfurters and luncheon meats. The gel structure provides a matrix for retaining moisture and fats and gives chewiness to the products. No meat proteins are required to form the gel structure; a process has been described for making simulated sausages of the frankfurter and bologna type in which all of the protein is supplied by soy isolate.[169]

Solutions containing 8% or more isolate gel on heating as described earlier; firm, resilient, and self-supporting gels are obtained at protein concentrations above 16 to 17%. Also studied was the influence on gelation of added salts, lipids, and polysaccharides commonly used in foods.[100] Further work, however, is necessary to learn how to control or enhance gelation in various food systems.

Several approaches have been taken to impart meat-like textures to soy flours and isolates.[171,178] In one method, aqueous slurries of flours or isolates are agitated vigorously at high temperatures. This treatment orients and coagulates the

46 *Soybeans as a Food Source*

proteins to form shredded masses that have meat-like textures.[170] This method has not been commercialized.

Another approach to meat-like textures is extrusion of soy flours. In this process (Figure 20) soy flour is mixed with water, flavor, colors, and other necessary additives, fed to a cooker-extruder, heated under pressure, and then extruded. Expansion of the product occurs on extrusion. Size and shape of the extruded mass is controlled by the arrangement of dies and the speed of a cutting knife. The final product is then dried, cooled, and packaged.[171]

A recent patent gives some of the preferred conditions for extrusion to form meat-like textures.[158] The soy flour should be low in fat (0.5% or less) and carbohydrate content should not exceed 35 to 40%. Addition of 5% starch causes excessive puffing with a loss of desired texture. Moisture content preferably is 30 to 40%, but if sodium hydroxide is added to raise the pH, the optimum range is 23 to 34% moisture. The pH

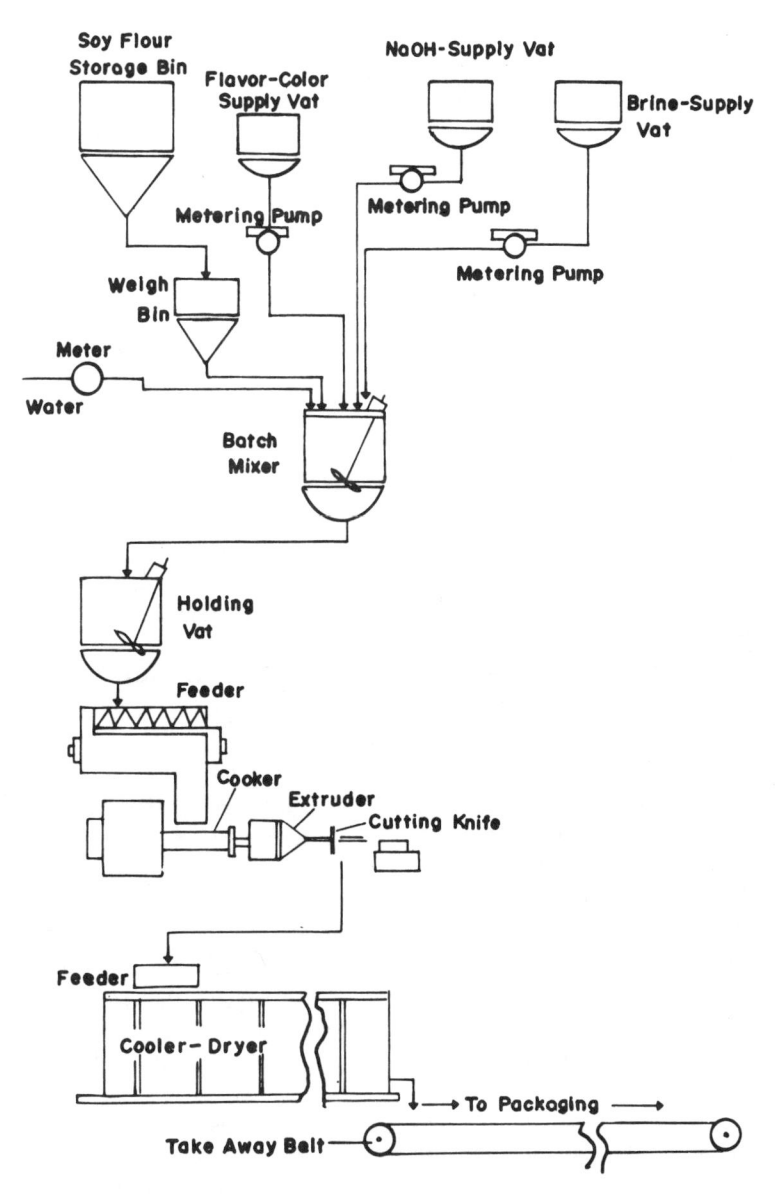

FIGURE 20. Flow sheet for cooking-extrusion to texturize soy flours. (From Ziemba, J. V., *Food Eng.*, 41(11), 72, 1969. With permission.)

range can be 6 to 9 but the preferred range is 7.5 to 8.7 and is obtained by adding an alkali such as sodium hydroxide. Although protein may vary from 30 to 75%, a content of 50% is most desirable. At protein levels above 75% a gummy product is obtained. Additives which are believed to cleave disulfide bonds in the protein molecules are recommended; 0.01 to 0.5% elemental sulfur, 0.04 to 2% potassium sulfide, or 0.1% sodium sulfide can be used. Temperatures during extrusion may range from 270 to 300°F, but 280°F is considered optimum. Pressures in the extruder vary from 300 to 600 psi.

The extruded products are crunchy when dry; thus, when properly flavored, they can be used to simulate fruits, nuts, or vegetables.[165] On hydration they are fibrous and chewy. Expanded products can also be prepared by heating a doughlike mass of soy flour in an autoclave and then releasing the pressure rapidly.[139]

A more complex and more expensive process than agitation-coagulation or extrusion is spinning of protein fibers from protein isolates (Figure 21). A continuous process has been developed whereby the proteins are isolated from soybean flakes and used directly, thereby eliminating drying costs and permitting greater control of the physical properties of the proteins used.[171,179] After extraction, acid-precipitation, and washing the protein is re-slurried in water and metered to a mixing pump where alkali is added to form an alkaline spinning dope. The filtered dope is then forced through spinnerets into an acid-salt coagulating bath to form filaments or fibers about 0.003 in. in diameter. Next, the fibers are washed and stretched to alter textures which can range from tender to tough by controlling the degree of stretching. The stretched fibers are then treated with binders to form bundles of fibers to which fat, flavor, color, stabilizers, and other ingredients are added. The order of treatment varies with the type of product being made, but after adding the binders and other ingredients, pressure is applied to form the simulated meat. After cutting into slices, cubes, bits, or granules (Figure 22) the product may be baked, broiled, cooked, or smoked. Textures can be varied from a crisp crunchiness typical of fried bacon to the chewiness of white chicken or turkey meat, and a number of applications for these products have been suggested.[172] Two concerns are producing these products on a commercial scale at present.[178]

An alternative method for making fibrous soy proteins consists of extracting defatted flour with water containing 0.1% sodium sulfite (to prevent formation of disulfide bonds) and precipitating the proteins with acid at pH 4.6 to 4.9. The curd is then mixed with salt and water and extruded into a water bath at 20°C. The extruded fibers are heated in a water bath at 90°C for 1 min to make them tough and elastic. The fibrous material is claimed to be useful as a hamburger extender.[180]

Meat-like fibers can also be made by mixing egg albumin, casein, or defatted soy flour with sodium alginate and extruding the mixture into a coagulating bath containing an acidic solution of calcium acetate. A calcium alginate gel fiber forms which acts as a carrier for the proteinaceous material. The gel imparts much of the texture, but if the protein is heat coagulable, it also contributes.[181]

The ancient process of making dried tofu (also known as kori tofu) in Japan imparts an interesting texture to soy proteins, but this technique appears to have received little attention in the U.S. In this process, soybean milk made from soaked soybeans is coagulated with a calcium salt to form tofu which is then ground, washed, pressed into blocks, sliced, and frozen at -10°C. After aging at -1 to -2°C for two to three weeks, it is thawed to yield a porous, spongelike structure which is then washed, dried, and sold in the retail markets. On rehydration and cooking, dried tofu swells and softens but retains considerable texture.[130,174] A recent patent describes a modification of the dried tofu process to make fibrous, spongelike structures from soy protein isolates.[173] Defatted flakes are soaked in water and ground to a slurry. After removing the insoluble polysaccharides by filtering or centrifuging, the proteins are precipitated by acidifying or by adding a divalent cation such as calcium. The protein curd is washed and dispersed in water by grinding and then frozen at -7°C for six hours. Minute ice crystals result and a spongelike protein structure is claimed to form which is stable to subsequent thawing and washing. Use of the textured product as a meat-like material is suggested.

Dough Formation

Soy proteins in their three basic forms will form "doughlike" masses when mixed with limited amounts of water. These doughs, however, do not have the elastic and cohesive properties typical of

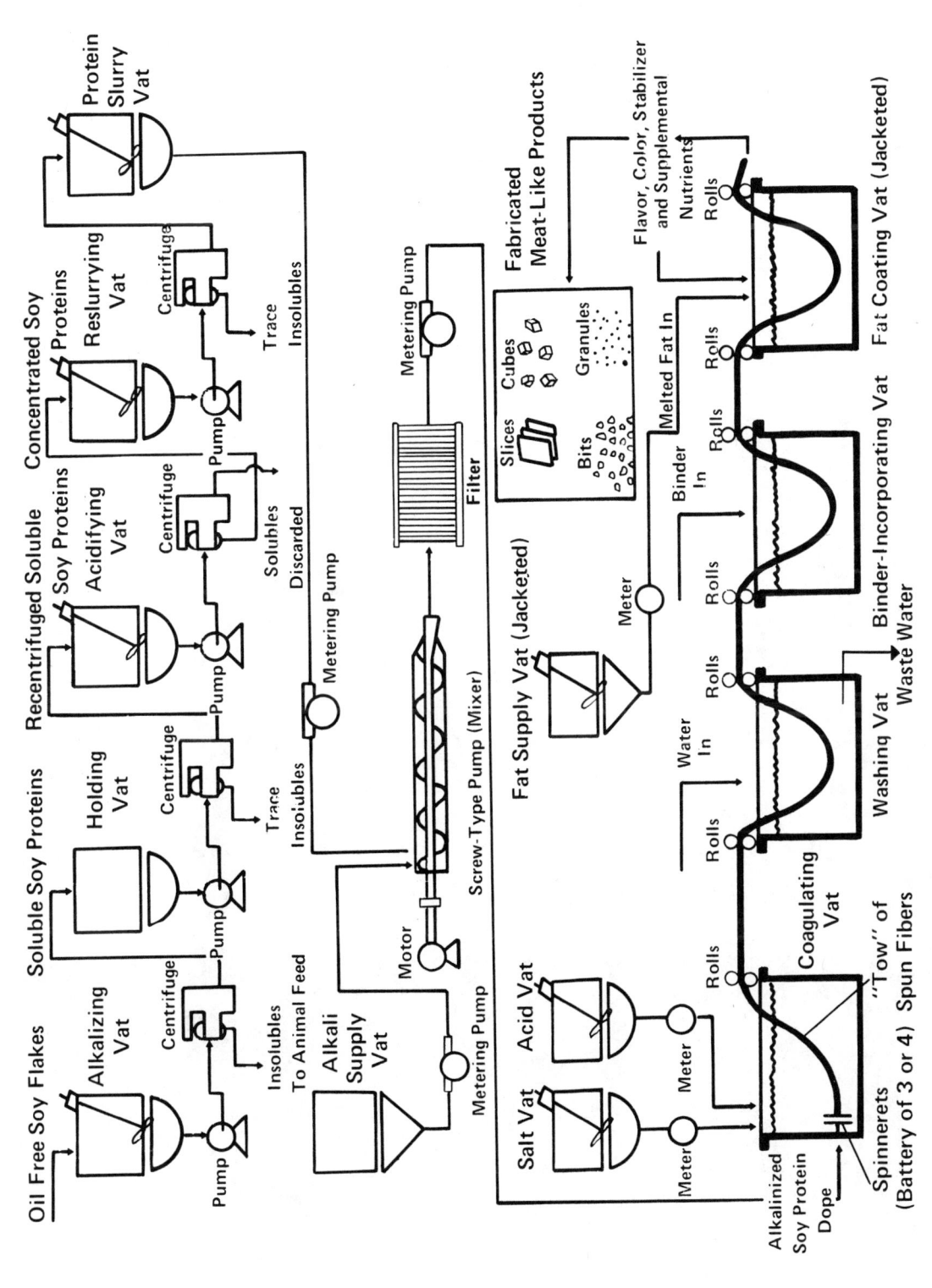

FIGURE 21. Flow sheet for manufacture of simulated meats by fiber-spinning of protein isolates. (From Ziemba, J. V., *Food Eng.*, 41(11), 72, 1969. With permission.)

the gluten proteins in wheat flour. Untoasted, defatted soy flour forms "doughs" when mixed with 40 to 60% water. Addition of less water gives a crumbly mass while more water results in a slurry.[139]

Adhesion, Cohesion, and Elasticity

These functional properties have been attributed to soy proteins in a variety of foods and are based largely on subjective observations instead of objective measurements. Difficulties in making and interpreting measurements of such textural properties were reviewed recently.[182] These properties, however, are often easy to demonstrate qualitatively. For example, the adhesive properties of isolates are used to hold together chopped chicken or turkey meat when they are pressed into "logs" or rolls.[159,166] Chemically treated soy flour added to macaroni reduces water absorption and thereby helps retain cohesiveness and elasticity of this product during cooking.[168] Elasticity and cohesiveness are observed in gels such as occur in frankfurters and bologna. Dried tofu after hydration also exhibits these properties.

Film Formation

The ability of soy proteins to form films is a desirable property in frankfurters and bologna. When soy flour-water doughs are autoclaved, a film forms on the surface. This surface film acts as

FIGURE 22. Forms that spun fibers (top) can be converted into include (clockwise) meat-like nuggets, ham-type cubes, poultry meat extender, red meat extender, simulated ground beef granules, bacon chip analog, and (center) chicken-type cubes. (From Worthington Foods, Inc. With permission.)

a barrier to water and aqueous solvents and is broken by grinding or slicing when the doughs are washed to remove water-soluble sugars, ash, and flavor components to produce protein concentrates.[139]

When meat is shredded and mixed with a combination of soy protein isolate or concentrate and egg albumin, the proteins coat and facilitate drying of the meat fibers. The protein coating retards loss of flavor in the dried meat particles, assists in rehydration, and provides desirable texture in the rehydrated product.[183]

Color Control

Soy flours are useful in controlling color in foods by two methods: the flours may act as bleaching agents or promote color formation in baked products. Raw soy flours are used as bleaching agents in white breads. Lipoxygenases in the soy flours oxidize polyunsaturated fats and the oxidized fats presumably bleach the wheat flour carotenoids, which are yellow, to a colorless form. A bread with a whiter crumb results. Recent work indicates that the classical soybean lipoxygenase does not bleach carotene; another enzyme which is heat sensitive is implicated.[184]

Soy flours contribute to color of baked products. In bread, crust color is reported to be enhanced; this effect is attributed to reactions between the soy protein and the wheat flour carbohydrates.[161] When soy flour is used in breading mixes, pancakes, or waffles it improves browning characteristics as well as retarding fat absorption during frying.

Aeration

Soy proteins are surface active; hence, they foam when whipped and are used in whipped toppings and frozen desserts. Pepsin hydrolyzates of soy proteins serve as whipping agents in confections, chiffon mixes, and angel food cake mixes. In contrast to the unmodified soy proteins, these hydrolyzates are soluble in the isoelectric range (pH 4 to 5). The pepsin hydrolyzates foam readily but depend on small amounts of unhydrolyzed protein[99] and additives such as corn syrups for stability of the foams.

Although protein isolates foam when whipped, the foams are not very stable because of foam inhibitors which can be removed by extracting the isolates with aqueous alcohols. After extraction with alcohol the isolates form very stable foams.[144,147] The foam inhibitors are probably residual lipids which have been isolated from alcohol extracts of isolates.[145] The foams obtained from alcohol-washed isolates, however, do not have the heat coagulation properties characteristic of egg albumin; alcohol-washed isolates cannot substitute for egg whites in angel food cakes.[147]

Nutritional Properties

Over 50 years ago Osborne and Mendel found that rats grew poorly when fed raw soybean meal and that dry heat did not improve the nutritional value of the meal.[185] Rats grew normally, however, when the meal was cooked on a steam bath for 3 hr. In the past 50 years a vast literature has developed on the nutritive properties of soybean protein, but moist heat is still used to improve the nutritional quality of soybean protein products for feeds and foods. Despite numerous experiments, there is still no clear-cut explanation for the poor nutritive value of raw soybean meal or the beneficial effects of moist heat on soybean meal.

The literature on this subject is often confusing and contradictory; two recent reviews give concise summaries of pertinent work for the last 30 years.[186,187] Alleged antinutritional factors and protein quality, therefore, are discussed only briefly.

Antinutritional Factors

Since moist heat readily inactivates the antigrowth factors of raw soybean meal, many workers believe that the factors are proteins such as trypsin inhibitors and hemagglutinins. Nonprotein components such as saponins have been suggested as antinutritional factors, but recent work does not support this view.

Trypsin Inhibitors—Five or more trypsin inhibitors are reported for soybeans[187] but only two—the Kunitz[188] and the Bowman-Birk[189] inhibitors—have been purified and studied in detail. Raw soybean meal contains 1.4% Kunitz inhibitor and 0.6% Bowman-Birk inhibitor.[190] Properties of the two inhibitors are summarized in Table 31. Both have isoelectric points in the acid range, but the Kunitz inhibitor has a molecular weight nearly three times that of the Bowman-Birk inhibitor. Another major difference is in the number of cystine residues per molecule. The seven disulfides in Bowman-Birk inhibitor

TABLE 31

Properties of Kunitz and Bowman-Birk Trypsin Inhibitors

Property	Kunitz*	Bowman-Birk*
Isoelectric point	4.5[188]	4.2[192]
Molecular weight	21,500[79]	7,975[191]
Amino acid residues	197[193]	72[191]
Cystine residues/mole	2[193]	7[191]
Stability to heat, acid, or pepsin	Unstable[188]	Stable[189]
Inhibition of chymotrypsin	Low[188]	High[189]
Pancreatic hypertrophy	+	+

*Superscripts are literature references.

apparently stabilize its molecular structure and make it resistant to denaturation by heat and acid or digestion by pepsin, whereas the Kunitz inhibitor is more readily inactivated by these agents. Although chymotrypsin is only weakly inhibited by the Kunitz inhibitor, it is strongly inhibited by the Bowman-Birk inhibitor. Both inhibitors cause enlargement (hypertrophy) of the pancreas in rats and chicks[190] but neither apparently has this effect on pancreas of calves[194] or swine.[195] Growth depression by ingestion of trypsin inhibitors has been attributed to endogenous loss of essential amino acids in the enzymes secreted by the hyperactive pancreas in response to the stimulatory effect of the inhibitors.[196] Liberation of a hormone-like factor by trypsin inhibitor has been suggested as a mechanism for stimulation of pancreatic secretion.[197,198]

Although both inhibitors are active against bovine trypsin, the Kunitz inhibitor has a low activity against the esterase activity of human trypsin.[199] The activity of human trypsin, however, is inhibited to a significant extent by Kunitz inhibitor when casein is used as a substrate to measure proteolytic activity.[200] It is not known whether ingestion of the inhibitors affects the pancreas in humans.

From a practical standpoint, trypsin inhibitors do not appear to be a serious problem in feeds and foods since they are largely inactivated by moist heat. Conditions of heating—time, temperature, moisture content, and particle size—influence the rate and extent of trypsin inhibitor inactivation. For example, atmospheric steaming (100°C) inactivates more than 95% of the trypsin inhibitor activity of raw, defatted soybean flakes in 15 min (Figure 23). Protein efficiency shows an accompanying increase in this same time, and flakes of 19% moisture gave a higher protein efficiency ratio than flakes of 5% moisture. In contrast, steaming whole soybeans, chips, or cotyledons for 20 min only partially inactivated trypsin inhibitors, apparently because of the large particle size.[202]

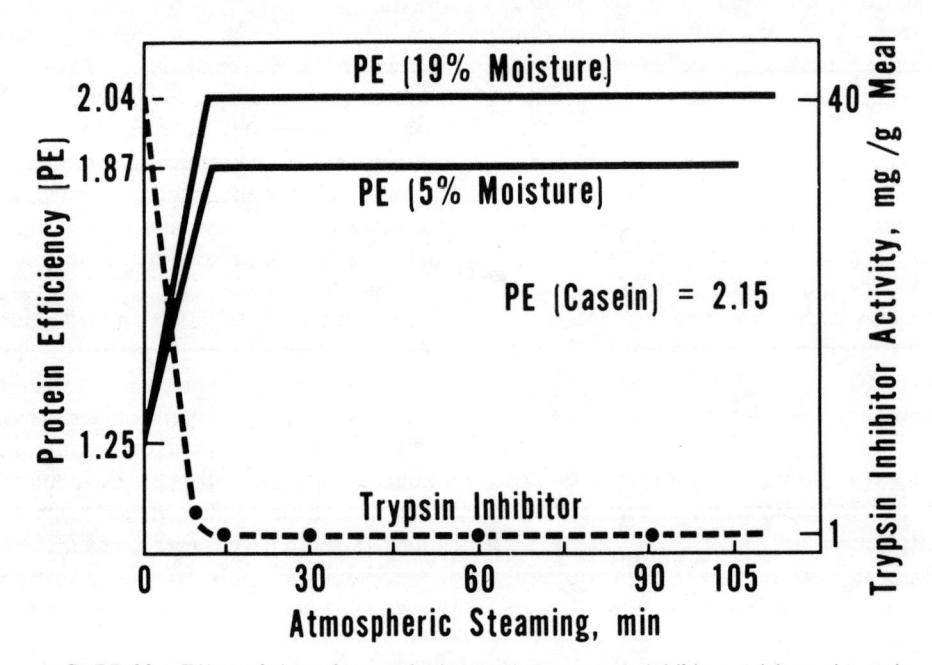

FIGURE 23. Effect of time of atmospheric steaming on trypsin inhibitor activity and protein efficiency ratio of raw soybean meal. (From Rackis.[201])

Atmospheric steaming inactivates most of the trypsin inhibitor in whole soybeans in 15 min if the initial moisture content is 20%. If the beans are soaked in water overnight to 60% moisture, 5 min in boiling water is sufficient to inactivate the inhibitors.[203] Further data on extent of heat treatment required to inactivate trypsin inhibitors are summarized elsewhere.[187] Small but measurable trypsin inhibitor activity can often be detected after heating; the known stability of Bowman-Birk inhibitor suggests that the residual inhibitor may be of this type. Measurements of residual chymotrypsin inhibitor activity would clarify this point because the Bowman-Birk inhibitor is a strong inhibitor of chymotrypsin.

A recent study reports trypsin inhibitor activity in a commercial protein isolate but no inhibitor was detected in canned frankfurters containing 1.5% isolate.[204] The heat treatment during canning inactivated the residual inhibitor.

Many of the conclusions drawn from studies on Kunitz inhibitor must be viewed with some reservation because of the heterogeneity of certain commercial preparations even when crystallized five times.[205] The possibility that a protein impurity or a tightly bound nonprotein impurity is responsible for some of the biological properties of the inhibitor has received slight consideration until recently.[206]

Hemagglutinins—Soybeans contain at least four proteins capable of causing clumping of red blood cells (hemagglutination) of rabbits and rats in in vitro tests.[86] These proteins are designated hemagglutinins; similar proteins are found in many legumes.[207] Defatted soy flour contains about 3% hemagglutinins.[208] The major hemagglutinin in soybeans has been isolated and characterized. It is a glycoprotein containing 4.5% mannose and 1% glucosamine, has a molecular weight of 110,000, and appears to contain two polypeptide chains.[81] The ability of hemagglutinins to cause clumping of red blood cells in a test tube serves as a useful assay procedure,[209] but there is no evidence that agglutination of red cells occurs when hemagglutinins are ingested. Hemagglutinin is readily inactivated by pepsin; thus, it probably does not survive passage through the stomach.[210] Furthermore, undigested hemagglutinin would have to be absorbed from the intestine to come into contact with red blood cells, an occurrence which seems unlikely because of the high molecular weight of the hemagglutinin.

Intraperitoneal injection of hemagglutinin kills rats, but the mechanism of toxicity is unknown.[211] Purified hemagglutinin fed at 1% of a diet containing heated soybean meal inhibited growth of rats but at a level accounting for only about one half of the growth inhibition with raw soybean meal. Growth inhibition was accompanied by a decrease in food consumption, which decline suggests a problem in palatability or an appetite depressant.[212]

Soybean whey proteins fed to rats on a heated soybean meal diet showed only one third of the growth inhibition obtained with raw soybean meal.[213] The conclusion from this study that hemagglutinin is unimportant as a growth inhibitor of rats and chicks by raw meal is open to question. Although whey proteins were added to the heated soybean meal diet at a level equivalent to that in raw meal, calculations show the diet contained only 53% of the hemagglutinin content of raw meal. One half of the hemagglutinin activity of raw meal apparently was lost during fractionation. Clearly, further studies are needed to assess the antigrowth properties of hemagglutinins.

Black bean hemagglutinin decreases food absorption and nitrogen retention in rats. It is suggested that hemagglutinins combine with intestinal cells and thus interfere with absorption of nutrients.[214] This hypothesis has not been tested with soybean hemagglutinins.

Soybean hemagglutinins are readily inactivated by heat (Figure 24) and inactivation is complete when maximum growth response is obtained. Hemagglutinins thus present no known problems in foods if preparation includes proper heating of the soy ingredient at some step of processing.

Soybean Saponins—Saponins are complex glycosides of triterpenoid alcohols and occur in soybeans to the extent of 0.5%. Because of their polarity, the saponins are insoluble in hexane and remain in defatted meal; defatted meal contains 0.6% saponins.[215] Although antinutritional properties have been ascribed to soybean saponins, recent studies show them to be harmless when ingested by chicks, rats, and mice at 0.5 to 3% of the diet. At the highest level the saponin content was about threefold higher than in a 50% soybean meal-supplemented diet.[216] Neither saponins nor sapogenins were found in blood of rats, mice, or chicks kept on diets containing 20% soybean meal. Thus, the saponins are not absorbed; they remain intact until they leave the small intestine but are hydrolyzed by bacterial enzymes in the cecum and colon. The saponins inhibit various enzymes including cholinesterase and chymotrypsin but inhibition is not specific. Soy proteins and other dietary proteins will also bind saponins.[217] Approximately 0.4% saponins were obtained from a laboratory preparation of soy protein isolate;[145] when isolates were heated in dilute acid solutions crystalline, but apparently modified, saponins were obtained.[146,218] The effect of interaction of the saponins with soy proteins is still unknown. The saponins are an extremely complex mixture and only limited separations have been obtained to date.[218,219] A thorough review of the chemical and biological properties of soybean saponins is available.[220]

Unidentified Factors — Some workers believe that trypsin inhibitors and hemagglutinins do not contribute to the antinutritional properties of raw soybean meal.[206] Other workers who take a less extreme view, however, can account for only a part of the antigrowth effects of raw meal in terms of trypsin inhibitors[201] and hemagglutinins.[212] Consequently, there still appear to be unidentified antigrowth factors in raw soybean meal. A recent study reports isolation from raw soybean meal of a growth inhibitor of mice. Soybean whey (Figure 19) contained most of the growth inhibitor when the meal was fractionated. In one experiment the whey was separated into three fractions by gel filtration on Sephadex G-50. The first fraction contained only traces of trypsin inhibitor but presumably hemagglutinins were present. When fed to mice it caused a small increase in pancreas size and a small decrease in growth rate, but neither effect was statistically significant. Fraction two, which contained the trypsin inhibitors, caused enlargement of the pancreas and inhibited growth. Fraction three, the last to elute from the column, was free of trypsin inhibitors, did not cause pancreatic hypertrophy, but inhibited growth.[206]

In a second experiment, fractions one and two were fed together at one half of the trypsin inhibitor level used in the first study when fraction two was fed. Pancreatic hypertrophy occurred, but no growth inhibition was apparent. It is not clear whether growth inhibition would have occurred at higher levels of combined fractions one and two, or if the growth inhibitor occurs mainly in fraction three and was incompletely separated from fraction two in the first experiment. Fraction three, containing the growth inhibitor, was a complex mixture of compounds having molecular weights less than 5000. The growth inhibitor was dialyzable and stable to mild acid hydrolysis at

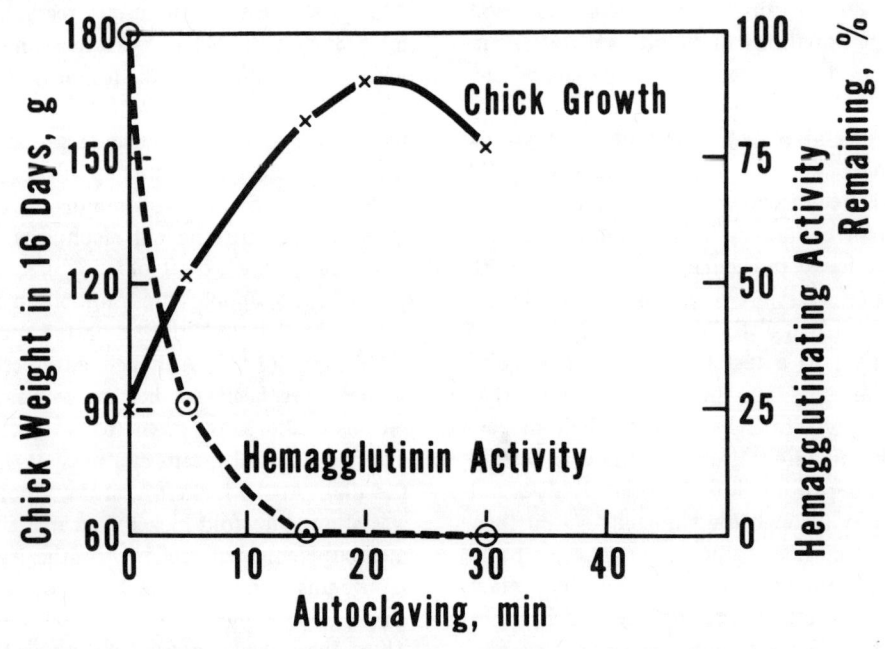

FIGURE 24. Effect of time of autoclaving raw soybean meal on hemagglutinin content and growth response with chicks. (Data from Liener.[209])

100°C; the latter result is surprising because of the common belief that antigrowth factors in meal are inactivated by autoclaving. Possibly toasted meal does not give the maximum growth response potentially present. These results show that unidentified antinutritional factors still exist in raw soybean meal.[206]

Protein Quality of Soybean Products

Until the 1960's information on nutritive value of soybean protein was largely limited to defatted flakes, meals, and flours. Moreover, most of the studies were concerned with use of soybean meal as an animal feed. Since commercial introduction of concentrates and isolates in 1959 and their increasing use in foods, these fractions have received considerably more attention. Studies with humans, however, are still limited.

Quality of soy protein products depends on several factors: (a) amino acid composition; (b) presence of antinutritional factors; (c) digestibility; (d) overall composition of the diet; and (e) nutrient requirements of species involved. Items a, b, and c are of primary importance in considering the various soy protein forms as protein sources. In the preparation of isolates, for example, fractionation occurs; this results in a change in amino acid composition as well as in removal of the antinutritional factors occurring in the whey. Items d and e are of greater importance when a specific food is being considered, i.e., an infant food, dietary item, or a snack food. Nutritional requirements for an infant differ greatly from the needs of an adult who may be trying to lose weight.

Amino Acid Balance—The essential amino acid contents of soy protein types are given in Table 32. Also included is the amino acid pattern for hen's egg protein recommended as a reference protein of good nutritional quality by the FAO-WHO Expert Group. Most of the amino acid levels in the soy proteins are equal to or exceed the levels in egg proteins with one exception. The sulfur amino acids are low, and as a result the protein scores for the soy proteins are low as compared to egg proteins. The lower score for isolate as compared to flour and concentrate results from loss of amino acids in the whey proteins during isolation. It is therefore necessary to supplement with methionine when isolates are the sole source of protein as is being done with infant formulas.[222] Alternatively, soy proteins can be blended with other proteins to provide a good balance of essential amino acids. For example, cereal proteins which are low in lysine can be blended with soy proteins to make mixtures which are better than either protein source by itself.[223] Problems of formulating such blends were discussed recently.[224]

Soy Grits and Flours—The superiority of heated versus raw soybean meal for feeding non-ruminants is well established, and conditions for producing meals of optimum nutritional value by toasting are carefully applied by the soy industry. Toasted flakes and flours, however, lack the functional properties found in products with less heat treatment. Consequently, for many food uses grits and flours of less than optimum nutritional value are used but heat treatment during processing of the food product is relied upon to develop maximum nutrition. The relationship between extent of heat treatment and nutritive value as measured with rats is seen in Table 33. The need to toast grits and flours for human use is not known, but autoclaved soybean protein in the form of whole beans, defatted flour, and a flour based infant formula has a biological value comparable to egg white and liver protein as measured by nitrogen balance studies in adult humans.[225] Feeding of raw and autoclaved soy flours to adult humans showed both capable of supporting a positive nitrogen balance and maintaining body weight. Feeding autoclaved flour, however, resulted in about a 20% greater retention of nitrogen than with raw flour. Both flours caused flatulence, but no other gastrointestinal symptoms were observed.[226]

Full-fat soy flour blended with rice flour and a commercial soy-based (full-fat flour) infant food were compared with milk powder in a 6-month study with 28 infants in Taiwan. Protein efficiency ratios for the soy flour-rice flour blends were lower than PER values for the commercial product and milk powder, but the differences were not significant. Infants fed the soy products gained weight and height at rates comparable to infants on the milk diet.[227]

In another study, soy flour-based infant formulas were fed to young pigs after a two-week pretesting during which they were fed either a low or high milk protein formula. The pigs on the low-protein pretest diet showed the soy flour formulas to be poorer than a formula based on soy isolate supplemented with methionine. Pigs on the high-protein pretest diet showed the same weight gains on either isolate or flour formulas, although the latter still showed lower PER values. The high-protein prefeeding apparently enabled the pigs to store protein which tended to minimize differences in protein quality between diets in the subsequent four-week feeding period.[228]

Textured Soy Flours—By proper control of pressure, temperature, and moisture during extrusion, textured flour products can be adequately processed to inactivate antinutritional factors.[134] Reduction of lysine and methionine contents in textured flours suggests that such products are more likely to be overcooked than undercooked.[229] Overheating full-fat flours by autoclaving caused a decrease in lysine and cystine, but only the loss of cystine had an adverse effect on the PER.[230] A four-week feeding trial with mice (Table 34) showed the textured flours to have PER values 74 to 81% of those for casein. Studies with humans show that textured soy flour maintained adult males in positive nitrogen balance for 15 days. Supplementation with methionine caused a decrease in nitrogen balance while addition of lysine to the textured flour diet caused a small increase in positive nitrogen balance.[229]

Concentrates—Early studies showed that com-

TABLE 32

Essential Amino Acid Patterns for Soy and Hen's Egg Proteins

Amino acid	Flour*	Concentrate†	Isolate†	Egg proteins‡
Isoleucine	119	115	121	129
Leucine	181	188	194	172
Lysine	161	151	152	125
Total "aromatic" A.A.	209	220	227	195
Phenylalanine	117	125	134	114
Tyrosine	91	95	93	81
Total sulfer A.A.	74	73	60	107
Cystine	37	40	34	46
Methionine	37	33	27	61
Threonine	101	100	93	99
Tryptophan	30	36	34	31
Valine	126	118	120	141
Protein score‡	68	68	56	100

*Calculated from data of Rackis et al.[126]
†Calculated from data in Technical Service Bulletins, Central Soya Co., Inc.[127,128]
‡From FAO-WHO Expert Group Report.[221]
#Based on total sulfur-containing amino acids.

TABLE 33

Protein Solubility and Nutritive Value of Full-fat and Defatted Soy Flours

	Defatted*		Full-fat†	
Heat treatment	PDI	Relative protein efficiency	NSI	PER
Negligible	90–95	40–50	–	–
Light	70–80	50–60	–	–
Moderate	35–45	75–80	50	1.82
Toasting	8–20	85–90	21	2.15

*From Horan.[131] Dried skim milk equals 100% efficiency.
†From Mustakas et al.[134] PER for casein = 2.5.

TABLE 34

Protein Efficiency Ratios of Textured Soy Flours

Protein source	Corrected PER	Relative efficiency
Casein	2.50	100.0
Unflavored textured flour	2.03	81.2
Beef flavored textured flour	1.93	77.2
Ham flavored textured flour	1.84	73.6
Casein + unflavored textured flour (1:1)	2.19	87.7

(From Hamdy et al.[231])

TABLE 35

Protein Efficiency Ratios of Commercial Protein Concentrates

	PER Methionine supplementation	
Concentrate	None	0.15%
A	2.29	3.00
B	2.16	2.88
C	2.36	3.06
Casein	2.50	

(From Meyer.[140])

TABLE 36

Protein Efficiency Ratios of Soy Concentrate: White Bread Blends

Concentrate:bread protein ratio	PER
0:100	1.1
25:75	1.8
50:50	2.5
75:25	2.6
100:0	2.4
Casein	2.8

(From Wilding et al.[223])

mercial concentrates had low PER values unless they were heated.[232] More recent work on three commercial concentrates indicates that PER values are 87 to 94% of the PER for casein (Table 35) and that heat treatment was unnecessary as a result of improved manufacturing practices. When supplemented with 0.15% methionine, all concentrates out-performed casein.

Another recent study confirms the high nutritive value of one type of soy protein concentrate (prepared by heat treatment followed by water leach, Figure 18) and demonstrates the value of concentrates as a supplement for wheat flour (Table 36). The PER for the concentrate was 86% of the value for casein and a maximum PER was obtained when concentrate supplied 75% of the protein and wheat flour furnished the remainder.

Isolates—Nutritive properties of isolates are variable (Table 37) and probably reflect variations such as varietal differences in soybeans and in processing conditions, including NSI value of the starting flakes and the amount of heat used during extraction and drying. Isolates (preparations A-D, Table 37) containing 0.09 to 1.86% trypsin inhibitor caused small but insignificant increases in pancreas weights in rats, and PER values ranged from 1.4 to 1.8. Toasted soybean meal gave a PER of 2.0; fractionation changes the distribution of essential amino acids in isolates as compared to meal (Table 32). Heating decreased the trypsin inhibitor content of isolate D but did not improve its PER significantly. Isolates I to IV had PER values of 1.4 to 1.9 as compared to a PER of 3.0 for casein. Isolate I showed no improvement on autoclaving but gave a marked response when supplemented with methionine hydroxy analog. In contrast, isolates II to IV gave increases in PER when heated. A much larger increase in PER on heating was noted in a study on isolates used in making spun fiber for simulated meats.[234] Apparently some isolates may be free of growth inhibitors but are marginal in sulfur amino acids while others may be less deficient in sulfur amino acids but contain heat labile antinutritional factors. Variability in isolates has also been noted by solubility measurements and ultracentrifugal analyses.[99]

A recent study indicates that some commercial isolates contain 3.3 to 4.5% trypsin inhibitor and that varying amounts of residual trypsin inhibitor persist in infant formulas made from isolates. Results with three infant formulas showed one to be higher in trypsin inhibitor and lower in supporting growth of rats than the other two.[222] Feeding of isolate-based infant formula supplemented with methionine to pigs showed the supplemented soy protein to have a protein quality approximately 85% of milk protein.[228]

Most isolates are prepared with a controlled amount of heat treatment to maintain high solubility. Conditions of protein extraction and precipitation are thus important in limiting the amounts of biologically active factors such as trypsin

TABLE 37

Protein Efficiency Ratios for Isolates

Isolate preparation*	Trypsin inhibitor content (mg/100 g protein)	Pancreas wt	PER Unheated	PER Heated	References
A	450	0.51	1.82	–	
B	129	0.52	1.65	–	
C	93	0.51	1.76	–	Rackis et al.[233]
D	1,855	0.56	1.40	–	
D (Heated)	86	0.45	–	1.63	
Casein	–	0.43–0.47	2.25–2.29	–	
I	–	–	1.36	1.46	
II	–	–	1.41	2.27	
III	–	–	1.77	2.29	Longenecker et al.[232]
IV	–	–	1.91	2.11	
I + 0.7% MHA†	–	–	–	2.24	
Casein	–	–	3.00	–	

*Designations of authors. All isolates were commercial preparation except D, a laboratory sample.
†Methionine hydroxy analog.

inhibitor found in the isolates. Little information is available concerning this point, but spray drying causes a decrease in trypsin inhibitor content of isolates as compared to freeze drying. Precipitation of isolates with calcium ion from hot (90 to 100°C) solution also decreases trypsin inhibitor content, but a small inhibitor activity (Bowman-Birk inhibitor?) remains.[235]

Increasing levels of isolate added to bread cause a progressive improvement in PER; bread containing 10% isolate had a PER of 2.26 as compared to 2.98 for the casein control.[176]

Spun Protein Fiber Foods—Simulated meat products made from isolates have been evaluated nutritionally in rats, dogs, children, and adult humans. A simulated beef granule containing 28.8% soy fiber, 12.3% egg albumin, 11.8% wheat gluten, and 9.6% soy flour gave PER values lower than for casein but as good as with dried beef. Nitrogen balance studies in children showed no important differences between the simulated beef product and skim milk at a protein intake of 2 g/kg/day. The starting isolate had a surprisingly low PER but responded very favorably when heated; residual heat labile antigrowth factors appeared to be present. These factors were partially washed out or inactivated during spinning since the fiber had a much higher PER than the isolate.[234]

Positive nitrogen balance was maintained for 24 weeks in adult humans on a diet containing spun fiber foods as the protein source. The subjects remained in good health throughout the study, although a few complained of abdominal cramping and flatulence. Maintenance of good health was confirmed by clinical and laboratory determinations.[236]

Oriental Foods—Although of lesser interest to Western food technologists, nutritional data on Oriental soybean foods are available. Examples selected from the literature are shown in Table 38. Many of these foods have been consumed for centuries and approach or surpass casein in protein quality.

TABLE 38

Protein Efficiency Ratios for Oriental Soybean Foods

Food	PER	References
Tofu (soybean curd)	1.93	
Natto (fermented soybeans)	1.00–1.52	
Edamane (green soybeans)	2.27–2.41	Standal[237]
Soybean sprouts	1.36	
Casein	1.44	
Soybean milk	2.11	
Curd*	2.20	Hackler et al.[238]
Casein	2.86	
Tempeh	2.48	
Casein	2.81	Smith et al.[239]

*Precipitated with acid from soybean milk.

Foods Containing Soy Proteins

Many of the uses of soy proteins have been alluded to in discussing their functional properties. Oriental food products are reviewed briefly since they are a significant part of the overall use of soybeans and a portion of the U.S. soybean crop is exported for this purpose. Several manufacturers make Oriental foods such as tofu and offer them for sale primarily in the large cities on the Pacific coast and in Hawaii where large numbers of people of Oriental descent live. Such sales of tofu were estimated to be about $3 million in 1969.[129]

Oriental Foods

Traditional soybean products in the Orient can be conveniently divided into two classes: non-fermented and fermented. The fermented types were reviewed recently.[240] Here Japanese foods will be given major attention since more information is available about them and several are produced commercially in Japan.[130,174] Table 39 lists amounts of soybean foods produced in Japan in 1967.

Tofu—Bean curd or fresh tofu is the major soybean food in Japan. It is prepared by water soaking, grinding, and cooking beans and filtering off the insoluble residue to obtain soy milk. A protein-oil curd is then precipitated by adding calcium sulfate to the warm milk. After the supernatant or whey is removed the curd is carefully washed, sliced, and sold. Tofu has a soft, gelatinous texture and is perishable. It is made fresh daily in over 40,000 plants in Japan. About one third of the tofu production is deep-fat fried to make aburage. Defatted meal is sometimes mixed with whole soybeans to prepare tofu.

A new development is "packaged" tofu which is made by putting soybean milk into a polyethylene or polyvinylidene chloride bag with a coagulant, sealing, and sterilizing. The coagulant may be calcium sulfate, gluconodelta lactone, or lactide—the latter two compounds hydrolyze to their respective acids, gluconic and lactic acids, which coagulate the protein by lowering the pH. A more concentrated soy milk is used than in preparation of regular tofu, and the whey is not removed.[130,241]

Dried (Kori) Tofu—This product is made from soybean milk by precipitating the curd with calcium chloride to form a hard and grainy tofu. After removing the whey, the curd is ground, shaped, washed, and sliced. The slices are rapidly frozen at -10°C and then aged at -1 to -2°C for 2 to 3 weeks. During aging the product becomes porous and spongy. After aging the tofu is thawed, dried, and packaged. Some types of tofu are treated with gaseous ammonia before packaging. Ammonia treatment makes the product swell more and softer when it is subsequently cooked in hot water by the housewife. Dried tofu has a longer shelf life than fresh tofu but is subject to browning and oxidative rancidity at high temperatures and high humidities. It is made on a large scale in about 40 factories in Japan.[130]

Kinako—This non-fermented product is made by dry roasting soybeans, dehulling, and then grinding. It is used as a cake base, and when mixed with sugar it is used on baked rice cakes.

Miso—A fermented food, miso, is made by cooking soybeans, blending with koji (steamed rice covered with a growth of *Aspergillus oryzae*), salt, and water and then inoculating with a yeast. Next, the mixture is fermented for several months to a paste-like consistency. Miso is used to flavor soups and vegetables. A variety of types of miso is produced based on color of the finished product, ingredients used, and geographical region of production. Most miso is prepared from whole beans.

Natto—Another popular fermented soybean product in Japan is made by inoculating cooked soybeans with *Bacillus natto* and incubating for 14 to 40 hr at 40°C. Fermentation covers the cooked beans with a viscous sticky substance which forms

TABLE 39

Production of Soybean Products in Japan for 1967

Food	Production (1,000 metric tons)
Whole soybean products	
Tofu and fried tofu	295
Miso	169
Natto	47
Kori-tofu (dried or frozen tofu)	34
Shoyu (soy sauce)	15
Kinako	12
Others	70
	Total 642
Defatted soybean products	
Shoyu	154
Tofu and fried tofu	77
Miso	8
Others	45
	Total 284

(From Watanabe.[130])

long stringy threads when the product is pulled apart. Natto is eaten with cooked rice seasoned with soy sauce or salt.

Shoyu (Soy Sauce)—This seasoning is well known in the U.S. but the traditional Japanese product is made by fermentation whereas the product most U.S. consumers are familiar with is prepared by acid hydolysis and has a different flavor than fermented soy sauce. Cooked defatted soybean flakes are mixed with roasted wheat and inoculated with *Aspergillus oryzae*. After the mold grows for 45 to 65 hr, salt solution is added and the fermentation proceeds for 8 to 12 months. The liquid portion is then separated from the insoluble residue, pasteurized, filtered, and bottled. One of the large Japanese producers exports its product to the U.S., and it is found on the shelves of many supermarkets.

Tempeh—This is another fermented product, but it is used in Indonesia rather than Japan. Soybeans are soaked overnight, dehulled, cooked, and mixed with a previous batch of tempeh. After incubating for 24 hr or less, *Rhizopus oligosporus* and related organisms grow and bind the soybeans together into a cake-like mass in which individual soybean cotyledons are still readily detected. The raw product is sliced and then fried. It serves as a main dish rather than as a flavoring for other foods as is true of many of the Japanese fermented products.

Domestic Foods

Most domestic uses of soybean proteins are for their functional properties, as discussed earlier, but with increasing emphasis on nutrition it is likely that more soy products will be used as supplements for foods deficient in protein quantity and quality. New foods now under development, such as the simulated meats, will undoubtedly increase the amounts of soy protein used.

A list of food products containing soy proteins is shown in Table 40. No attempt has been made to make the list complete; rather its purpose is to point out the variety of present applications. Many of the products are available in the retail markets whereas several are food ingredients.

Baked Goods—Although concentrates and isolates have been experimentally incorporated into bread and other baked goods,[176,223] defatted and full-fat soy flours are the primary soy protein forms used by the baking industry. The flours are low in price (Table 26) and are often used at low levels for their functional effects. Thus, flavor of these cruder soy proteins is not a problem. Of the 60 million pounds of soy flours consumed by the baking trade (Table 27), 6 to 7 million pounds are of the defatted, enzyme active type which is used as a bleaching agent in white breads. Use is limited to 0.5% (based on weight of wheat flour) by the Standards of Identity for white bread. In England, enzyme-active flour is of the full-fat variety and it is estimated that 90 to 95% of the bread produced there contains soy flour.[155,161]

Many bakers in recent years have begun using soy flours as a regular ingredient in their breads and also as a partial replacment for non-fat dry milk which has been rising in cost as a result of decreasing milk production in 1965-69 and because of increases in price support levels. Wholesale prices for non-fat dry milk are about 28 cents per pound; a 17% price increase occurred in 1970. Use of non-fat dry milk in baked goods, consequently, has fallen off sharply. One company is now marketing a soy product designated as "special process" soy flour with properties intermediate to those of a soy flour and a protein concentrate. This product containing 60% protein is used as a partial or complete substitute for non-fat dry milk or whey solids in bread. Bread baked with this soy product has a PER as good as or slightly better than bread containing non-fat dry milk.[167] A baker in a major city is using this special soy flour to bake Kosher bread and rolls. The soy flour is claimed to provide the functional advantages of using milk products and also complies with religious dietary codes. The 60% protein product can also be used as a replacement for non-fat dry milk in commercial cake flours.[177] Because of its economic advantages, use of this soy flour can be expected to increase in the future.

A related development is the blending of sweet cheese whey with soy flour[151] to prepare products containing protein levels approximating non-fat dry milk. These whey-soy blends are being manufactured by at least three companies and are being promoted for use in baked products such as breads, doughnuts, cakes, and cookies. Cheese whey-soy flours blends are quoted at 19 to 21 cents per pound as compared to 28 cents per pound for non-fat dry milk.

Products designed to duplicate whole milk powders have also been introduced commercially. They consist mainly of isolated soy protein, vegetable fats, and low dextrose equivalent corn

TABLE 40

Commercial Food Products Containing Soy Proteins

Company	Protein form used	Product
Archer-Daniels-Midland Co.	Defatted flour	TVP® (textured vegetable protein-simulated meats)
Borden, Inc.	Isolate	Bacon Tasters® (snack)
	Isolate	Neo-Mull-Soy® (infant formula)
Continental Baking Co.	Defatted flour	Hostess doughnuts
Deltown Chemurgic Corp.	Isolate	Dellac 25-28 SX (whole milk powder replacement)
Estee Candy Co.	Defatted flour	Dietetic cookies
		Dietetic candy
Far-Mar-Co., Inc.	Defatted flour	Ultra-Soy® (textured vegetable protein-simulated meats)
General Foods Corp.	Defatted flour	Fortified Oat Flaks
General Mills, Inc.	Isolate	Bac*Os® (fried bacon analog)
		Bontrae® (meat analogs)
	Pepsin hydrolyzed protein	Angel Food Cake Mix
Gerber Products Co.	Concentrate	Beef Stew (infant food)
		Chicken Stew (infant food)
	Defatted flour	Cookies (infant food)
Griffith Laboratories, Inc.	Acid hydrolyzed protein	Vegamine® (meat-like flavoring agent)
H. J. Heinz Co.	Defatted flour	High Protein Cereal (infant food)
		Vegetables, Dumplings, Beef and Bacon (infant food)
Interstate Brands Corp.	Defatted flour	Hollywood Diet® (special formula bread)
Thomas J. Lipton, Inc.	Isolate	Chicken Baronet (casserole mix); Beef Stroganoff (casserole mix); Chicken La Scala (casserole mix); Ham Cheddarton (casserole mix)
Mead Johnson and Co.	Isolate	ProSobee® (infant formula)
McCormick and Co., Inc.	Defatted flour	Imitation Bacon Bits
Nestle	Acid hydrolyzed protein	Maggi® Hydrolyzed Plant Proteins
Pet, Inc.	Defatted flour	Sego® (dietary beverage)
Quaker Oats, Co.	Concentrate	Life® (breakfast cereal)
Ross Laboratories	Isolate	Similac Isomil® (infant formula)
Schutter Candy Co.	Flour, Isolate	Bit-O-Honey® (candy)
A. E. Staley Mfg. Co.	Defatted flour	Mira-Tex® (textured vegetable protein-simulated meats)
Swift and Co.	Defatted flour	Texgran® (textured vegetable protein-simulated meats)
	Defatted flour	Chili with beans
	Isolate	Veal with Vegetables (infant food)

TABLE 40 (continued)
Commercial Food Products Containing Soy Proteins

Company	Protein form used	Product
	Isolate	Beef with Vegetables (infant food)
	Isolate	Chicken with Vegetables (infant food)
	Defatted flour	Provide® (cheese whey-soy flour blend)
H. P. Taylor Co.	Defatted flour	Textrasoy® (textured vegetable protein-simulated meat)
Worthington Foods	Isolate	Stripple Zips® (fried bacon analog)
	Isolate	Wham® (ham analog)
	Isolate	Prosage® (sausage analog)
	Isolate	Stripples® (bacon analog)

syrup solids and are recommended for baking and other uses as replacements for whole milk.

Until recently the practical limit of soy flour use in bread was 4 to 5% of the weight of wheat flour. Work at Kansas State University now indicates that levels as high as 12% can be used if emulsifiers such as sodium stearoyl-2-lactylate, calcium stearoyl-2-lactylate, and ethoxylated monoglyceride are incorporated into the dough. These emulsifiers or dough conditioners enable bakers to obtain normal loaf volumes and the bread is reported to have the texture, aroma, and taste of traditional bread. Protein can be increased 50% higher than in normal bread, and lysine content can be doubled.[242] With recent emphasis on better nutrition in the U.S., this could develop into a sizable market if implemented. Several special-formula breads containing low levels of soy flour are already on the market.

Soy flours are commonly used in doughnuts for their functional effects—control of fat absorption. Soy flours are also used in other baked goods such as cakes, cake mixes, cookies, and pancake mixes. In England, for example, the majority of the commercially produced cakes contain full-fat soy flour.[161]

A high-protein base for baked goods was described recently.[243] For preparation of high-protein cookies the base contained 40% ground oats, 29% soy protein concentrate, and 8% soy protein isolate plus non-fat dry milk, egg solids, and methionine. Cookies made from this base had an essential amino acid content equivalent to whole egg protein. A formulation containing concentrate and isolate for use in high-protein breads was also given.

Meat Products—The second largest use of soy proteins is in meat products; detailed information on this subject can be found in a recent review.[156] Nearly 40 million pounds of flours and grits are utilized as extenders and as functional ingredients in processed meats. Soy flours are added to cooked sausages and nonspecific loaves but have the disadvantages of changing mouth feel of the products and having residual soy flavor which is undesirable. Soy grits are added to sausages, meat loaves, patties, and chilies. Grits do not affect mouth feel as flours do, but they also are limited in their use because of flavor. In addition to the flavor problems of soy flour and grits which tend to make their use self-limiting, USDA regulations restrict the amounts of these materials that can be added to various meat products (Table 41).

Concentrates and isolates have found increasing usage in the meat industry because of their lower flavor intensities and as a result of increasing prices

TABLE 41

Levels of Soy Products Permitted in Meat Products

Meat product	Soy product	Permitted level, %
Fresh and cooked sausage	Flour or grits	3.5
	Concentrate	3.5
	Isolate	2.0
Chili con carne	Flour or grits	8.0
	Concentrate	8.0
	Isolate	8.0
Spaghetti with meat balls	Flour or grits	12.0
	Concentrate	12.0
	Isolate	12.0

Based on regulations of Meat Inspection Program, USDA.

for non-fat dry milk since 1965. Because concentrates and isolates are high in protein, their addition to meat products does not tend to lower the protein content of finished meat products as occurs when cereal products are added. Concentrates and isolates are used in products including beef patties, nonspecific loaves, cooked sausages, chilies, meat balls, and chicken and turkey rolls. Use of isolates in Federally inspected plants requires that the isolates contain 0.1% titanium dioxide as a tagging agent for analytical monitoring of the amount of isolate in a finished product. Titanium is determined spectrophotometrically after ashing.[156]

One company is mixing isolate with chopped beef and chicken to make dehydrated meats for soup and casserole mixes. Addition of the isolate facilitates dehydration of the meat product; circulating-air oven drying can be used instead of freeze-drying which is more expensive.[183]

Simulated Meats—This application of soy proteins is still in its infancy, but in 1969 sales of all simulated meats (meat, poultry, and seafood analogs) were estimated to be about $10 million. One prediction is that sales by 1980 will be about $2 billion but will account for less than 5% of the total market for meat, poultry, and seafoods.[152] A more optimistic prediction is that by 1980 sales of meat analogs and related products will equal meat sales.[178]

Extruded soy flour products have a cost advantage over the spun fiber items but still contain residual flavors plus the oligosaccharides-stachyose and raffinose-which appear to cause flatus when soy flour is ingested. Extruded soy flour products are being made by at least five companies at this time. Prices for the textured soy flours range from 27 to 80 cents per pound, depending on flavor, color, density, shape, and quantity. In use, the dry products are hydrated with 2.5 to 3.0 parts of water. Thus, cost on a ready-to-serve basis can be as low as 8.5 cents per pound.

Because the meat-like extruded flours are dry, they are easily stored and have good shelf-life.

Bacon-flavored textured flours are sold in the retail stores and are marketed to hotels, restaurants, and institutions. Beef- and chicken-like products have been developed for addition to canned chili, dry-mix chicken a la king, Sloppy Joe mix, and Spanish rice. Other uses are replacement of part of the freeze-dried meats in dry soup mixes and simulation of meat particles in dips, crackers, and snacks. Fine particles of textured flour are used as meat conditioners in beef and pork patties where they reduce stickiness of meats and thereby permit better release from patty-making equipment.

Two companies are manufacturing and marketing spun fiber meat analogs. One company has spent more than 300 man years over a period of 8 years in research and development and in 1970 completed a multimillion-dollar plant for production of the spun fiber materials.[152,153] The analog plant is adjacent to a new convenience food plant; thus, much of their spun fiber production will be for internal use. Prior to construction of the new facility, this company produced about one million pounds per year of frozen analogs and twice as much of the dried materials. Present production of both dry and frozen analogs is several times greater.[153] A fried bacon bit analog has been on the retail market for several years but has not been advertised nationally because of limited production facilities until recently.

Much of the meat analog production is being designed for the food service industry—hotels, restaurants, armed services, and institutions—which makes up a sizable segment of the food market. Nearly 40% of all meals consumed in the U.S. are eaten outside of the home. Scarcity of skilled help in the food service industry places a premium on convenience in food preparation. Meat analogs and prepared foods containing them that merely require thawing or reconstitution and heating are being developed, and predictions are that significant amounts will be used during the next decade.[152]

Analogs of beef, ham, chicken, and seafood have been formulated and tested for wholesomeness and acceptability by prison inmates and university students. Most test subjects preferred the simulated beef and chicken over the ham and seafood analogs.[236] Further testing has been conducted by hotels, restaurants, and state institutions. Quality and acceptance can be expected to improve, however, as more experience is gained and better flavors become available. Good meatlike flavors were unavailable prior to the development of the meat analogs.

A strip-like bacon analog was recently test marketed. The analog is make by randomly laying down spun isolate fibers and holding them together with an edible binder. Alternating layers of red and uncolored fibers are employed to simulate the lean and fat portions of bacon. After shaping, heat setting, and slicing, the product is frozen.[244] For preparation the product is merely heated. No cooking or frying is necessary; consequently there is no shrinkage. Bacon shrinks to one fourth of its weight on being broiled or fried.

The bacon-strip analog was test marketed in Fort Wayne, Indiana, from September 1968 through February 1969.[245] During the first three months the analog was intensively promoted with emphasis on its low calorie and cholesterol content as well as its economy as compared to bacon. Sales of analog to bacon reached a ratio of 1:24 during the first phase. Advertising was reduced during the second 6 months and analog-to-bacon sales dropped to 1:79. A strong promotional campaign would thus have to be maintained to achieve a high level of sales. The study did indicate, however, that some persons buying the analog were not using bacon for dieting or health reasons. The analog is still being sold in Fort Wayne and in at least two other cities where it was introduced in 1969-70.

Wholesale prices of the spun fiber products range from $.75 to $.85 per lb for the frozen forms and $.85 to $1.35 per lb for the dry types.

Breakfast Cereals—Soy flour is added to breakfast cereals to improve the amino acid balance of the cereal proteins and to increase protein content. One manufacturer is adding soy flour to an oat-flake cereal; the product contains 18% protein. Another is adding soy protein concentrate and sodium caseinate to oat flour to raise the protein content of a ready-to-eat cereal to 18%. A number of high-protein cereals are on the market, but their protein quality often leaves much to be desired because of processing or protein composition of the starting materials. Four commercial oat cereals with protein contents of 15.5 to 23.0% had PER values of 2.18 to 2.79, whereas a rice cereal with 21.3% protein and a wheat cereal with 19.5% protein had respective PER values of only 0.40 and 0.19.[246] Supplementation of cereals with soy flours or concentrates is thus a potential area of increased use of soy proteins.

Infant Foods—Examination of labels on products on supermarket shelves reveals that many infant foods now contain soy proteins in their various forms. A high (35%) protein cereal, for example, contains defatted soy flour as the major ingredient. One company uses soy flour in their bottled vegetable and meat foods, a second uses protein concentrates, and a third adds isolate to their formulations. Cookies containing soy flour are also available for infants (Table 40).

Infant formulas based on soy protein are manufactured by five companies in the U.S. These foods are designed for infants that are allergic to cow's milk. One product is based on soybean solids extracted from ground soybeans (soy milk), two products are formulated from soy flour, and three formulas contain isolates fortified with methionine.[222,228] Another formula is made from isolate without methionine supplementation and is also recommended for adults preferring a vegetarian diet.

Beverages—A variety of beverages containing soy proteins has been developed in recent years. Some of these beverages are designed for developing countries; problems of flavor, texture, acceptability, costs, and product stability have been encountered. These and other protein-fortified foods were reviewed recently.[247]

One company has made an isolate-containing beverage for the vegetarian market for several years. At least two companies have recently formulated soy-based powders as replacements for whole milk powder. Ingredients include soy isolate, vegetable oil, corn syrup solids, vitamins, and minerals. Suggested uses include reconstitution to beverages, but we are unaware of the success of this application. Greater usage will probably be as a replacement for milk in processed foods.

Imitation milks received much attention and were selling well in certain areas several years ago, but sales have declined again. Improvement in quality plus a greater economic advantage over cow's milk seem necessary before these products become successful. This application, however, represents a large potential market for soy proteins if problems of flavor, nutrition, and legal restricitons can be overcome.[248]

Dietary Foods—These specialty items command relatively small shelf space in the supermarkets, but a number of them contain soy proteins. Cookies and candies containing soy flour are available. A low-cholesterol egg powder containing soy isolate is being offered. Calorie-controlled beverages which were very popular several years ago contained soy protein. Several have been reformulated and soy protein, particularly flour, has been reduced in quantity or eliminated because of flavor and flatus problems.

Snack Foods—Snack items have taken over an increasing amount of shelf space in supermarkets and play a major role in the diet of many Americans. Snack sales were under $2 billion in 1959 but reached over $3 billion in 1968—an increase of over 50% in a decade. These foods, traditionally high in carbohydrates and fat, have been criticized as contributing to nutritional problems in several segments of our population.

Many of the snacks on the market today are typical fabricated foods made by extrusion of starch and flour doughs, but the conventional ingredients cannot be fortified merely by adding protein products. Expansion of the snack during extrusion may be reduced or eliminated by protein additives. Expandable starches that are compatible with protein ingredients are now available, however, and several soy protein processors have formulated snack prototypes fortified with flours, concentrates, and isolates. It is anticipated that protein-fortified snacks will appear on the market in the near future. A bacon-flavored fabricated chip containing isolate appeared on the market in 1969, but the protein content of this snack item is not known.

Miscellaneous Uses—Soybean proteins are insoluble in their isoelectric range (pH 4 to 5); hence they cannot be used in food products of this pH if solubility is needed to give a desired functional property. Insolubility of the proteins in the isoelectric pH region is eliminated by hydrolyzing the proteins with pepsin. Hydrolyzed proteins are utilized as foaming agents for syrups used as nougat, fudge, and creams in candies; for chiffon pies; and in whipped cake mixes. These hydrolyzates are also added to angel food cake mixes to increase volume of the egg white proteins on whipping. Pepsin-modified proteins have typical hydrolyzate flavors—bitter and salty. Bitterness results from formation of peptides containing leucine in the carboxyl terminal position.[249] Pepsin-modified proteins have been available for about 20 years; use in 1967 was estimated at less than a million pounds.[149]

Other uses of soy proteins (flours and grits) include acid hydrolyzates for soy sauce and meat-like flavors, foam stabilizers in beer, carriers for artificial spices, to reduce stickiness in manufacture of confections, supplementation of pasta and macaroni products, and addition to grease applied to baking pans to aid browning of breads and cakes in areas contacting the pans. Meat-like flavors (referred to as hydrolyzed vegetable proteins) are made by hydrolyzing soy concentrate with hydrochloric acid until 35 to 58% of the total nitrogen is converted to alpha amino nitrogen or 54 to 89% of the peptide linkages are broken. The hydrolyzate is then neutralized to pH 4.5 to 7.0 with sodium hydroxide and spray dried.[250] These hydrolyzates give meat-like flavors to bland materials added to meats as extenders or binders and to simulated meats. Recent studies identified many of the flavor compounds found in hydrolyzed vegetable proteins.[251,252]

Problem Areas

Use of soy proteins in foods has increased dramatically in the last decade and many of the present developments have not yet reached their full potential. Nonetheless, the industry is faced with a number of problems that will tend to retard development unless they are solved. These problems include flavor, flatus factors, functional properties, nutritive value, legal aspects, and disposal of waste products.

Flavor—The first generation soy products, flours and grits, suffered serious defects in flavor which were partially corrected by controlled application of moist heat (toasting) to eliminate the characteristic beany and bitter flavors of raw soy. Toasting, however, develops a typical flavor that is incompatible with many foods and seriously alters functional properties such as protein solubility.

Isolates and concentrates, the second generation of soy products, were developed to lower flavor levels and also to reduce the amount of non-protein material found in flours and grits. Recent taste panel evaluation of flours, concentrates, and isolates showed that beany and bitter flavors still occur in concentrates and isolates.[253] Reduction in flavor level is apparent, but there are still a number of uses where these residual flavors are objectionable.

Several laboratories have reported on compounds believed responsible for flavors of soybeans but, as in most studies of flavor constituents, it is difficult to tell which compounds are the most important. A basic question is: Do the flavor compounds preexist in the soybean or are they formed by enzymatic and nonenzymatic reactions when the seed is crushed?

In a study of the volatile compounds lost during moisture determinations on soybeans, a sample of whole beans was heated in a vacuum at 150°C.[254] Analyses of the volatile materials revealed only the normal gases found in air plus ethanol and hydrogen sulfide. Concentration of ethanol found was only 1/20 of the amount reported by others for full-fat meal.[255] Likewise, no acetaldehyde was found whereas 0.8 ppm is reported for ground soybeans.[255] While these results suggest that crushing of the seed is necessary to generate the numerous compounds isolated from meals and flours, the purpose of the study was to identify the major volatiles lost on drying rather than flavor constituents. Some destruction of flavor compounds may have occurred at the high temperature used to prepare the volatile fraction.

In a systematic study of flavor components of soybeans the following classes of compounds were isolated: carbonyl compounds,[256] phenolic acids,[257] volatile fatty acids and amines,[258] and alcohols and esters.[255] Acetaldehyde, acetone, and n-hexanal were the major carbonyl compounds isolated. n-Hexanal has a low flavor threshold and is believed to contribute to the beany flavor. The carbonyls were isolated from water slurries of full-fat and defatted meal; thus, oxidation of polyunsaturated fatty acids by lipoxygenase is highly likely. Breakdown of hydroperoxides, therefore, could account for formation of the carbonyl compounds. Hexanal, however, was also found in the volatile fraction from ground soybeans in the absence of added water.[255] Vacuum stripping of defatted flakes likewise yielded acetaldehyde, acetone, and n-hexanal, but flavor of the flakes was unchanged.[259] Since complete removal of these carbonyl compounds was not demonstrated, residual amounts may have remained in the stripped flakes. Binding of n-hexanal and n-hexanol to soybean globulins has been demonstrated.[260]

Of nine phenolic acids found in defatted flour, syringic acid was highest in concentration.[257] The sour, bitter, and astringent flavors of the phenolic acids may contribute to the flavor of defatted soybean flour. Two of the phenolic acids—p-coumaric acid and ferulic acid—are reported to be precursors of 4-vinylphenol and 4-vinylguaiacol which contribute to off-flavors when defatted soybean meal is autoclaved.[261]

Only small amounts of volatile fatty acids were found in full-fat meal but isocaproic, n-caproic, and n-caprylic acids were the major components. Ammonia, methylamine, dimethylamine, piperidine, and cadaverine comprised the volatile amine fraction.[258] The importance of the amines in the overall flavor problem is unknown. Seven alcohols were isolated from full-fat meal and, of these, 3-methyl-1-butanol, n-hexanol, and n-heptanol are thought to contribute to the beany flavor of soybeans.[255]

The flavor profile of conventionally prepared soybean milk is very complex and is largely caused by the action of lipoxygenase and possibly other enzymes on the lipids during soaking and wet grinding of the beans prior to cooking. Ethyl vinyl ketone (green, beany) and 1-octen-3-ol (mushroomy, earthy, musty) are reported to contribute significantly to the flavor of soybean milk.[262,263] The latter compound accumulates in soybeans during water soaking and is optically active. Thus, it appears to be formed by an enzymatic reaction. In a comprehensive study of soybean milk volatiles approximately 80 gas chromatographic peaks were found. Of these compounds 41 were positively identified and an additional 13 were tentatively identified. About 25% of the volatiles consisted of hexanal.[264] It is thus clear that lipoxygenase activity in full-fat products must be prevented if undesirable flavors are to be avoided or minimized.

Two methods have been reported for inactivating lipoxygenase during wet processing of soybeans. The first involves grinding unsoaked beans with water at temperatures between 80 and 100°C and holding at these temperatures for 10 min to inactivate the enzyme.[16] The second method consists of grinding soybeans in an acid slurry (pH

(3.8) followed by cooking on a steam bath and neutralizing to pH 6.5. This procedure is reported to yield a bland product. The low pH inhibits lipoxygenase which is then inactivated by heat. Head space analysis of a slurry of beans ground in acid revealed a mixture of methane, ethane, propane, butane, and acetaldehyde. In contrast, head-space vapors from a cooked bean slurry initially ground in water (pH 6.5) contained a large amount of acetaldehyde plus pentanal and hexanal, and the slurry had an oxidative off-flavor.[265]

At present we do not know whether lipoxygenase and other enzymes are responsible for flavors of defatted flakes and flours. Since these products are processed at low moisture contents, lipoxygenase activity would be expected to be small; however, the possibility that limited enzyme action can give rise to organoleptically significant amounts of flavor compounds cannot be ruled out. Moreover, defatted flakes and flours are not lipid-free; commercial products contain 0.5% or more of residual oil plus an additional 2.5% of bound lipids.[266] Thus, when defatted flakes are processed with water, as in the preparation of isolates and certain concentrates, there appears ample opportunity for lipoxygenase to act on the residual oil and bound lipids. Laboratory preparations of isolates contain lipoxygenase.[267]

The bound lipids are extracted from defatted flakes by solvents more polar than hexane. Hexane:ethanol azeotrope removes the bound lipids and also reduces the flavor level of hexane-defatted flakes.[266] Isolates prepared form hexane-ethanol extracted flakes have less flavor than isolates from hexane-defatted flakes.[268] The bound lipids appear to be combined with the proteins since isolates contain a similar fraction; the lipids in isolates are not removed by hexane but are extracted with aqueous ethanol.[145]

A better understanding of the nature of the flavor compounds and their origin is essential for increased food uses of soy protein products. The empirical approach of processing soy grits and flours with moist heat has been successful in developing a sizable market, but further expansion will be difficult until new information becomes available to serve as a basis for a more scientific approach to processing. Preparation of concentrates and isolates is based on information developed 20 to 30 years ago, but improvement in flavor of these products is still on an empirical basis.

Flatus Factors—Soybeans in common with many other legumes cause flatus when ingested. Tests with humans showed that the flatus-producing factors were not associated with the hulls, fat, water-insoluble polysaccharides, or protein. The flatus factors are high in defatted flour but are progressively reduced when flour is processed into concentrates and isolates. Isolates and sodium caseinate appeared to inhibit flatus production. Flatus factors were concentrated in the 80% ethanol-soluble fraction obtained in preparing concentrate and in the whey solids fractions resulting from the isolation of the protein.[269,270]

Further studies indicate that flatus is caused by fermentation of the low-molecular-weight sugars—raffinose and stachyose—which are not digested because humans do not possess α-galactosidase activity in their digestive tract. When soybean milks with low, medium and high levels of raffinose and stachyose were fed to rats, flatus production correlated with oligosaccharide level.[271] Moreover, addition of stachyose to a casein diet at a level equivalent to the stachyose content of soy milk produced a greater volume of flatus than soy milk. The soy milk with a low level of oligosaccharides was prepared by treating the milk with an α-galactosidase preparation. The possibility thus exists that the oligosaccharides can be removed by enzymatic procedures, but problems with browning due to formation of reducing sugars can be anticipated.

Flatulence resulting from ingested textured soy flour is reported, but quantitative data are lacking.[229] Studies at practical levels of consumption of the textured materials are needed to ascertain whether flatus problems will be a deterrent to large-scale use of these products. Extrusion of concentrates instead of flours is an alternative for solution of the flatus problem, although this will increase production costs.

Functional Properties—Properties of present soy protein products limit or exclude their use in certain food applications. For example, there is a great deal of interest in adding proteins to carbonated beverages. Soy proteins are not suited for this purpose because of their insolubility in such acid systems. The possibility of selective enzyme modification of the proteins to alter their solubility does not appear to have been explored in depth. Modification with pepsin as presently practiced is too severe and introduces flavor problems.[272] Moreover, factors such as phytate that influence solubility of the proteins in acid solution are not clearly understood.

Use of isolates in products such as coffee whiteners is limited because of "feathering," presumably as a result of insolubility at the pH of coffee. One approach to this problem is chemical modification. Reaction of proteins with acetic anhydride or other anhydrides acylates amino groups and lowers the isoelectric point, thereby making the protein more soluble when the pH is lowered.[273]

A better understanding of factors causing insolubility of isolates during the manufacturing process could result in more soluble products than are presently available. Insolubility results in undesirable settling out of protein in fluid products such as beverages.

Nutritional Value—Although soy protein have a good nutritional value when adequately processed, there are limitations when nutritional requirements are stringent. A good example is the use of soy isolates in infant formulas. Four formulas containing isolates are available in the U.S., but three are fortified with DL-methionine to compensate for the limiting level of sulfur-containing amino acid in the isolates.[222] Nutritive value and flavor are stumbling blocks to the development of imitation milks containing soy isolates. Proposed regulations of the Food and Drug Administration require that imitation milks contain proteins with a biological quality not less than 70% of that of casein.[274] Moreover, the amount and biological quality of the protein in an 8-fluid-ounce serving must be such that the quality of the protein expressed as a fraction of that of casein multiplied by the grams of protein is not less than 8.0. Lack of nutritional quality, therefore, must be balanced by increasing the quantity of protein.

Use of DL-methionine in imitation milks causes flavor problems; thus, alternative methods for supplementing soy isolates need to be tried. Blending with isolates from other sources is a possibility, but the number of proteins available for this purpose is very limited at present.

The importance of trypsin inhibitors in the diet needs to be established. Until such information is available, either fractionation or heat treatment must be used to eliminate these activities from food products. Heat treatment may be undesirable in some products because of insolubilization of protein; hence, the need to know whether it is necessary to inactivate the inhibitors when they occur in a given food.

Legal Aspects—Soy protein products used as food components are considered as food additives and must, therefore, be approved by the Food and Drug Administration (FDA) for safety and usefulness. The soy protein industry and FDA are presently working on a standard of identity for textured vegetable protein products. It is expected that nutritional quality of the products will be included in the definition and will cover both extruded flour and spun fiber types.

Because soy proteins are being added to foods such as meat items, which also have standards of identity, the amounts of soy that can be added are usually very limited (Table 41). Industry representatives feel that present and proposed USDA regulations are too restrictive and do not permit use of sufficient soy protein to effect improvements in nutrition and product quality and reduction in cost.[275] Clearly, future rulings by government agencies will affect growth of the soy protein industry.

Disposal of Waste Products—Production of concentrates and isolates results in waste products which must be disposed of, and disposal costs must be included in production costs. Concentrates yield whey-like by-products (Figure 18) which contain the soybean oligosaccharides sucrose, raffinose, and stachyose plus variable amounts of nitrogeneous constituents (depending on the method of preparation), flavor compounds, and ash. Isolate production yields whey plus the insoluble residue which consists mainly of the water-insoluble polysaccharides plus unextracted proteins. The solids content of the various "wheys" produced annually is estimated at 100 million pounds; about one half of the solids come from isolates and concentrates for edible use and one half derive from industrial isolate production.[276] The wheys have little economic value and one means of disposal is to concentrate them to a syrup of molasses-like consistency and color and then add them to animal feeds. Recovery of solvent is an economic necessity in the alcohol extraction process for making concentrates; thus, the solids are obtained as a normal operating step. Wheys from the acid leach process for concentrates and from isolates contain more protein than the alcohol leachings, but approximately one half of this protein can be heat coagulated and added back to the concentrate or isolate.[276]

Another approach to whey disposal is to use it to make up part of the moisture needed in toasting defatted flakes in manufacturing feeds. A foreign isolate plant reportedly is using this method.

CONCLUSIONS

Uses of soybean oil and protein have not reached the stage where "soybean foods" has become a household term, and this point may never be reached. Foods based primarily on soybeans are not likely to be developed or become popular. Instead, the trend is toward processed foods where the consumer becomes familiar only with the brand name of the product and is less aware of the ingredients. Soybean oil has already moved to the head of the list of ingredients in many products. In contrast, soybean proteins have only begun to appear somewhere in the middle of ingredient listing except for products such as simulated meats. As animal proteins continue to increase in price and as further improvements are made in the flavor of soy proteins, these materials will appear in increasing numbers of foods and in greater amounts than at present.

Improvement in the flavor of soy protein will probably be slow because identity of the compounds involved is still uncertain and present information suggests that a large number of compounds may be involved. Flours, concentrates, and isolates now available are prepared from hexane-defatted flakes which originally were a by-product of oil extraction and are now an excellent source of protein for feeds. The oil extraction process is efficient but may not be ideal for preparation of bland protein products. When the flavor problem is better understood, changes in present practices may be desirable. For example, if the flavor compounds arise during processing, it may be possible to add inhibitors, remove activators, change pH, or remove oxygen to prevent formation of the flavor compounds.

ADDENDUM

INTRODUCTION

Since publication of our original review in 1971, world food shortages have heightened interest in soybeans as a source of edible oil and protein. To keep this monograph up-to-date, we have reviewed the literature from 1971 through late 1974 and summarized the recent developments here. Several topics not discussed in the first edition have also been included to broaden coverage.

Origin of Soybeans

The often repeated statement that the earliest use of soybeans by man occurred in the 24th to 29th century B.C.[277] appears to be open to question. According to Hymowitz,[278] the 10th and 11th century B.C. or slightly earlier is a more suitable date based on archaeological findings and the earliest known use of the ancient Chinese symbol for soybean, shu. Accurate chronological dating in Chinese history prior to 840 B.C. appears to be most difficult. The supposed date of soybean use reported in the material medica of the legendary Emperor Shen Nung is now considered to be a fabrication of Han Dynasty (206 B.C. to 220 A.D.) historians. No traces of legumes have been found in any neolithic site in North China or records of oracles of the Shang Dynasty (about 1500 to 1000 B.C.). Indeed, most historians doubt if Emperor Shen Nung ever existed except in the minds of the Han historical writers.

Present evidence about the domestication of the soybean centers around analysis of the ancient character for soybean (shu) and its use in early writings. Pictorially it can be described as having the following:

1. A horizontal line symbolizing the earth;
2. Upper and lower lines depicting stem and roots;
3. Teardrop-like lines indicating nodules.

According to Hymowitz,[278] students of the ancient Chinese language believe this type of character for shu first appeared in records of the early part of the Chou dynasty (1000 to 220 B.C.) or no earlier than about the 11th century B.C. Of course, some time would be needed for the trial-and-error process of domestication to develop the soybean as a food source. Thus, food use probably originated in the latter part of the Shang dynasty. Evidence points to the provinces of Shantung, Honan, Kiangsu, and Anhwei in the North China plain as the site of domestication and not Manchuria, the major soybean-producing area for China today.

Soybean Situation — Future

Over the next ten years to 1985, soybean demand is expected to increase to 2.1 billion bushels, or nearly quadruple the 1960 level, and to be more than one billion bushels over the early 1970's level.[279] Domestic demand will increase about one half, but export will double. Present domestic use of soybean oil for food is over 7 billion lb and is projected to increase to over 10 billion lb by 1985. Meal use is expected to rise from over 12 million tons to 21 million tons.

The projections for soybean demand are based on predictions of increased consumption of fats and oils in the United States. These predictions are given in Table 42. Per capita consumption has increased for visible fats from 45 to 53 lb between 1960 and 1972, with soybean oil use increasing about 15 lb per person, from 36 to 58% of the total. By 1985 projected use is expected to increase to a total of 60 lb per capita, with soybean oil contributing 68% of the total of 15.5 billion lb. The recent trends in edible oil use are expected to continue, with salad and cooking oil consumption increasing over 5 lb per person. This outlet is most economical and attractive for refiners since salad oil will require less processing than shortening or margarine.

Soybeans face major competition as a source of fat from several oilseeds including peanuts, rapeseed, sunflower seed, and palm fruit. Palm is a perennial crop with two crops per year and is harvested throughout the year in Malaysia and Africa. It yields 1,000 to 4,000 lb more of oil per acre than any other plant. Production is projected to increase from 2.1 million metric tons in 1972 to 4.3 million metric tons in 1980.[280] Palm oil is now being refined in the United States on the Pafiic Coast in new facilities that eliminate alkali refining and reduce pollution.[281] Palm oil will compete with soybean oil for the shortening market that is projected to increase from 16.7 lb per capita to 19 lb in 1985. Palm oil use may also increase in margarine and cooking oils, but certain characteristics will have to be modified by proper processing to gain much of an outlet as a salad oil.

In the United States sunflowers are successfully grown in the Red River valley of the North, but insect resistance and improved varieties are needed in the southern areas before they will compete with soybeans. West Texas shows some promise for the future.[282] Sunflowers from other areas, particularly Russia, could again affect U.S. exports of soybeans to Europe. Rapeseed in Canada will increase in use and export, but probably not in the amounts needed to meet world demands and threaten soybean exports seriously. Rapeseed in Europe is expected to supply a portion of these projected needs. Peanuts from Africa have supplied much of the oil for Europe, but recent production has been held in many of the developing countries for food. Also, aflatoxin-producing organisms grow readily on peanuts in many of the Africa production areas, making the peanut less attractive for food or feed use. Soybeans contain inhibitors that slow growth of many mold organisms that grow on soybeans but do not produce much toxin.[13,283] The problems of energy supply and continuation of recent high prices for petroleum loom as big threats to increased soy demand and use.

TABLE 42

Food Fats Used and Soybeans Required for 1960, 1972, and 1985 Including Exports

Fat or oil	1960	1972	1985 Projection
	Pounds per capita		
Soybean	16	31	41
Other vegetable	10	10	9
All animal	19	13	10
Total use for food	45	53	60
Soybeans required	Millions of bushels		
Processing	406	750	1,100
Export	135	475	950
Total	541	1,225	2,050

(From *Fats and Oil Situation*.[279])

Demand for edible soy protein also increased in part from U.S. consumer reaction to high meat prices as well as the greater availability of textured protein foods. The first breakthrough in U.S. consumer adoption came in 1971 when USDA permitted use of textured protein products in the National School Lunch Program. In the spring of 1973 supermarkets began to offer ground beef-textured soy protein blends for sale in their meat departments. Many companies are now marketing textured products as meat extenders, and a few are offering vegetable protein ham-flavored slices, sausage-flavored links, and other meat-like products. The consumer has shown a willingness to try many of these products, and sufficient repeat sales are occurring to show that increased and continued sales will occur. Soybeans may become a major source of protein in any attempt to feed starving populations. Its acceptance by the people involved, as well as its cost, though low compared to many other sources, present problems that politicians and others have not solved. Transportation is both inadequate and costly. Population control needs more careful consideration and positive action for our human species to maintain peace and possibly to survive.

Recent Sources of Information

For the neophytes as well as the experts who wish either to learn about or review the recent advances in soybeans, several new comprehensive publications are available. They include the following books and proceedings of meetings:

1. Smith, A. K. and Circle, S. J., Eds., *Soybeans: Chemistry and Technology. Proteins*, Vol. 1, Avi Publishing, Westport, Conn., 1972, 470 pp.

2. Caldwell, B. E., Ed., *Soybeans: Improvement, Production and Uses,* Am. Soc. Agron., Madison, Wisc., 1973, 681 pp.

3. American Soybean Association, *Proc. World Soy Protein Conf.* Munich, Nov. 11 to 14, 1973, *J. Am. Oil Chem. Soc.,* 51, 49A, 1974.

4. *Soybeans: Production, Marketing and Use,* Bulletin Y-69, National Fertilizer Development Center, Tennessee Valley Authority (TVA), Muscle Shoals, Alabama, March 1974, 187 pp.

5. Altschul, A., Ed., *New Protein Foods. Technology*, Vol. 1A, Academic Press, N.Y. 1974, 511 pp.

The first of these publications is written for the chemist, engineer, and marketer who is concerned with the chemistry, processing, and sale of soybeans and their protein food products. Special attention is devoted to the genetic, chemical, and biological characteristics as they are related to the purification, processing, fermenting, and marketing of soy protein foods. Fourteen experts cover the subject matter in 12 chapters, and the book is edited by two scientists associated with soy protein for a combined total of more than 60 years.

The second book is primarily an up-to-date review of the factors involved in the improvement and production of soybeans. Chapters on morphology of the plant, speciation and cytogenetics, qualitative and quantitative genetics, and varietal development should serve as good sources for the plant geneticist seeking information basic to development of new superior varieties. Separate chapters on the effect and control of major soy pests, such as weeds, fungi, bacteria, viruses, nematodes, and insects, as well as on management of production with emphasis on water and mineral relations, nodulation, and photosynthesis complete the coverage related to improving production practices. Research on both approaches (i.e., superior varieties and production techniques) is needed to break an apparent yield barrier facing soybean producers. Short chapters on processing and products and the foreign and domestic markets for soybeans and their products complete the 20 chapters written by 34 subject-matter specialists.

The third publication gives the reader over 50 papers containing recent information that would be difficult and tedious to collect from other sources. Soy proteins, food products, their manufacture and properties, nutritive value, legal and regulatory aspects, and use in foods and assistance programs are among the subject areas included. These proceedings are, indeed, a valuable source of information for anyone vitally concerned with soy protein foods.

The collective paper-bound volume was published by the TVA and covers all aspects of soybeans as related to the southeastern area of the United States. It is a good source of information not only for this area but elsewhere in the world.

The fifth and most recent book covers the problems of technology as well as politics in attempting to supply malnourished populations with sufficient protein of adequate quality. It also covers new approaches that are being used in developed countries. Fourteen experts cover ten subject-matter areas including amino acid fortification of intact cereals and cereal products, protein enrichment of bread and other baked foods, legumes in foods, new approaches to marketing red meat and poultry, meat-like products from soybeans, new Japanese technologies, and legal regulation of protein foods for safety and establishment of quality.

In August 1975 the University of Illinois will sponsor an international meeting on soybeans. This conference and its proceedings should be of interest to all concerned with soybean production and use.

Soybean Organizations

A number of organizations in the United States have a primary concern with soybeans and their products. These include the American Soybean Association (ASA), American Soybean Association Research Foundation, Midsouth Soybean and Grain Shippers Association, National Soybean Processors Association (NSPA), and some 20 state growers' organizations. Both the American Oil Chemists' Society and the American Association of Cereal Chemists have contributed to the advancement of the chemistry of soybean products by their sponsorship of scientific meetings, symposia, and short courses. These societies have been leaders in publication of the papers presented on soybeans at their meetings as well as elsewhere. The American Society of Agronomy has served a similar function for the agronomy and genetics related to soybeans.

The ASA was founded in 1920, organized in 1925, and maintains offices in Hudson, Iowa. It publishes the *Soybean Digest*, of which one issue is the annual *Soybean· Blue Book*, a directory for industry. It authorized the American Soybean Association Research Foundation in 1965 which provides a means whereby firms and individuals who have a stake in the industry may contribute funds to implement research in fields related to soybeans. One major source of funds is the so-called check-off laws passed in the various states. For example, each farmer may contribute $0.005 per bushel when he sells his beans to an elevator or processor. A portion of these funds is assigned to specific promotion or research in each state, whereas another part is channeled to research or promotion through the Foundation. Between June 1972 and December 1974, for example, about $250,000 was contributed to the Foundation with about two thirds coming from the ASA and state growers' associations and one third from oilseed processors and related companies as well as concerns supplying material or equipment for the soybean industry.

The NSPA was organized in 1930 and Sheldon J. Hauck is the Executive Director with offices in Washington, D.C. It publishes a yearbook and trading rules. The National Soybean Crop Improvement Council, founded in 1941 by the processors, has as its Managing Director, Robert W. Judd of Urbana, Illinois. The Council cooperates with agronomy departments and experiment stations in soybean producing states, the U.S. Department of Agriculture, and all parts of the soybean industry to encourage research on production. The NSPA has a number of active technical committees including the Soybean Research Council.

The Food Protein Council was organized in 1971 and operates within the NSPA rules and policies to improve and promote soybean protein food products. It distributes information and assists in development of practical standards and tests.

Both the ASA and the NSPA work closely with the USDA Foreign Agricultural Service on market promotion. ASA concentrates on soybeans particularly for Europe, Japan, Taiwan, and Mexico, with offices in Brussels, Hamburg, Tokyo, Taipei, and Mexico City. NSPA promotes the marketing of oil and meal with offices in New Delhi and Karachi.

PRODUCTION

Short-Term Situation

Since 1971 the production of soybeans has increased to a record 1,566 million bushels in 1973, and Table 43 shows the USDA preliminary estimate for acreage, yield, and production for the ten leading states.[284] In 1974 adverse weather caused a reduction in acreage harvested and yield as well as total production, estimated at about 1,244 million bushels. The average price paid farmers rose from about $3.00 per bushel in 1971 to a high monthly average of $10.00 in June 1973. Prices continued high for the 1973 and 1974 crop years, averaging almost $6.00 per bushel for 1973,

TABLE 43

Soybean Production in the Ten Leading States in 1973 and 1974*

	Acreage, millions of acres harvested		Yield, bushels per acre		Production, millions of bushels	
State	1973	1974	1973	1974	1973	1974
Illinois	9.23	8.50	31.5	25.0	290.7	212.5
Iowa	7.90	7.07	34.0	28.0	268.6	198.0
Indiana	4.29	3.91	31.5	25.0	135.1	97.8
Minnesota	4.39	4.04	29.0	22.0	135.1	88.9
Missouri	4.70	4.35	27.0	24.0	126.9	104.4
Arkansas	4.65	4.30	25.0	20.0	116.3	86.0
Ohio	3.59	3.20	25.0	24.0	89.8	76.8
Mississippi	2.75	2.55	22.0	19.0	60.5	48.4
Louisiana	1.72	1.76	22.5	24.0	38.7	42.2
Tennessee	1.57	1.54	23.5	22.0	36.9	33.9†
Total – all states	56.41	52.51	27.8	23.7	1,566.5	1,243.9

*Preliminary.
†North Carolina may pass Tennessee in production in 1974.

(From *Fats and Oil Situation.*[284])

and fluctuated from $8.30 to $9.50 on the Chicago Board of Trade Futures Market according to the Wall Street Journal for October 23, 1974. Supply and disposition of soybeans as of November 1974 for 1973 to 1974 are shown in Table 44.

Since December 1972 strong demand for soybean oil for food use has continued despite increased prices. The average quarterly Decatur price rose from about $0.10 per lb in July to September 1972 to $0.44 per lb for the same months in 1974. Meal prices also rose from $104 per ton to $151 for the same periods, with some averages going as high as $310 as demand varied primarily because supply of protein meal from other sources (particularly Peruvian fish meal) fluctuated.[285]

The major increase in the use of soybean oil has been in the salad and cooking oil category. Greater use of liquid oil in salad dressings, widespread acceptance of hydrogenated-winterized soybean oil as salad and cooking oil for the home, and increased use of partially hydrogenated soybean oil for deep-fat frying of chicken, fish, potatoes, and other foods has led to over 3 billion lb being consumed in 1973 to 1974. Table 45 gives recent data on soybean oil consumption. The relation of soybean oil to other selected fats for food use is

TABLE 44

Soybeans: Supply and Disposition, 1973–1974

	Million bushels
Supply	
Carryover, Sept. 1	172
1974 Crop	1,244
Total	1,416
Disposition	
Processing	765
Exports	500
Seed, feed, and residual	91
Total	1,356
Ending stocks, Aug. 31	60

(From *Fats and Oils Situation.*[284])

shown in Table 46. Lately, increasing amounts of liquid soybean oil have gone into margarine with the major oil refining companies marketing their own "tub" margarines containing liquid soybean oil or manufacturing these "tub" margarines for supermarket chains. The total for soybean oil in all margarine amounts to about 1.5 billion lb. The impetus for this new outlet for unhydrogenated oil appears to come from higher costs for other oils and the consumer demand for polyunsaturates.

TABLE 45

Recent Use of Soybean Oil in United States

Billion of pounds

Year beginning October	Shortening	Margarine	Salad and cooking oils	Total
1970	2.08	1.38	2.29	5.78
1971	1.99	1.41	2.58	6.01
1972	2.10	1.49	2.62	6.23
1973	2.19	1.51	3.07	6.78

(From *Fats and Oils Situation*.[284])

TABLE 46

Consumption of Soybean Oil and Other Selected Fats for Food Use — 1973

Use in	Billions of pounds	% of total	Pounds per capita
Visible fats			
Margarine	1.49	79	7.1
Shortening	2.12	58	9.6
Salad and cooking oils	2.91	73	13.1
		% of visible fats	
Total (soybean)	6.52	58	29.8
Total all vegetable oils	8.90	79	42.4
Total animal fats	2.38	21	11.3
Total butter	0.81	7	3.9
		% of all fats	
Total all visible fats	11.30	43	53.7
Total invisible fats	15.43	57	71.8

(From *Fats and Oils Situation*.[286])

Labels now showing the amount of saturates and polyunsaturates in a serving of margarine are permitted and are found on an increasing number of brands. Several such labels show liquid soybean oil as the major ingredient, with 11 g of fat in 1 tablespoon of margarine (a serving). Such a serving contains 4 g of polyunsaturated acids and 2 g of saturated acids or a P/S ratio of 2. If such a ratio is desirable in all food fats, greater amounts of unhydrogenated or partially hydrogenated-winterized oil will probably have to come from soybeans.

Production of oil and meal is expected to decline in 1974 to 1975 with a total of about 765 million bushels to be crushed or 7% below the record 1973 to 1974 crush of 821 million bushels.[284] Processing capacity has increased to about 1.1 billion bushels and processing margins are expected to decline. Some of the increased capacity has come in the ability of the South to supply poultry feeders and foreign markets. Increased storage goes hand in hand with increased processing capacity.

Meal use increased to a record level of 13.8 million tons in 1973 to 1974, but a decline in the number of consuming animals is expected to drop domestic disappearance in 1974 to 1975 to about 12.6 million tons.

Storage and Exports

A new development may show the way for other foreign processors who wish to be assured of a steady supply. Through a market development program on advertising of identified soy oil, its use appears to be steadily increasing in Italy. After 12 months of promotion, Ferruzzi and Company's production of soy oil at two plants shifted to almost 100% label-identified soy oil. Previous production had been unidentified as to source of oil. Ferruzzi competitors began their own advertising program within a few weeks after Ferruzzi started his program.[287] In order to assure a steady supply and reduce the waiting period of his vessels at New Orleans, Ferruzzi found it highly desirable to operate his own elevators and river terminals in the Midwest and storage and loading facilities in New Orleans. His newest (1974) and largest Midwest river terminal at Davenport, Iowa can unload and store 150,000 bushels a day and deliver about the same amount to a barge in 4 hr. With terminals also at Pekin, Gulfport, and Beardstown, Illinois, Ferruzzi handles over 200 million bushels of corn and soybeans; this is the same as the quantity of soybeans the United States produced 30 years ago.[288]

Other storage and terminal facilities are steadily increasing to handle soybeans for domestic and foreign shipment. The lower than predicted acreage and yield for the 1974 crop year should have eased the demand for immediate new storage facilities, but they must be constructed and placed in operation to handle the increased production forecast for 1985.

Soybean Varieties

Until recently, most of the soybean varieties released for production were developed by the USDA and the State Experiment Stations, either jointly or separately. During the last five years, the number of varieties released by private concerns has grown rapidly. With about ten maturity zones from northern Minnesota to southern Louisiana, increased demands for soybean products, and the passage of the plant variety Protection Act (December 1970), the number of varieties available is likely to grow. In February 1974 the *Soybean Digest* listed ten comparatively new public varieties that were now more available, three new public varieties, and 46 relatively new private varieties.[289] Although the variety Wayne was grown on nearly 7 million acres or approximately one sixth the total acreage harvested for beans in 1971, it appears that continued varietal development to fill local needs will be required for some time. Importance of individual varieties is likely to diminish.

Yield Barrier

Soybean plants have an amazing ability to compensate. This trait helps to prevent low yields, but it also presents researchers with a so-called yield barrier. Whereas average soybean yields have increased only 6 bushels per acre since 1950, corn yields have almost doubled. The ability of soybeans to compensate appears to be responsible, and the limiting factor needs to be identified and the solution to increasing yields found. Examples of soybean's ability to compensate show how it differs from most other farm crops. If soybean plant populations are increased, yields are not increased because fewer seeds are obtained per plant. If the lower branches of the plants are removed, more matured beans can be harvested. Thus, combining weed control with destruction of the lower branches may give producers about an 8% increase in yields.

If nitrogen fertilizer is applied, the nitrogen-fixing bacteria that are symbiotic with the soybean plant produce less nitrogen. Nitrogen fixation by the soybean increases rapidly at flowering but slows down when the seeds are setting. Even application of nitrogen just before the seeds are laid down does not give much increased yield. One identified limiting factor appears to be photosynthesis. If the soybean plant is surrounded with a plastic barrier open at the top, carbon dioxide available to the plant is tripled and nitrogen fixation and soybean yields are almost doubled. Practical ways of achieving this greater use of carbon dioxide are being sought through research on growth regulators, development of nutrients, and other studies on the photosynthetic process in the soybean plant.[290]

Another hope that may prove to be very helpful in breeding soybean plants with improved characteristics and lead to a break in the yield barrier is the discovery of male-sterile character.[291] It appears to be effected by a single recessive gene pair. Use of the male-sterile plants permits natural rather than artificial crossing of lines, and large numbers of crosses are readily carried out with comparatively little effort.

Varieties and Antinutritional Factors

Stachyose and raffinose have been implicated as causative factors in digestive disturbances, such as flatus, caused by soybeans.[270] Hymowitz and co-workers[292,293] examined seeds from 78 varieties or lines of soybeans from Maturity Groups 00 through IV for total sugar, sucrose, raffinose, and stachyose, and part of their data is shown in Table 47. Considerable variation in sugar content occurred, and it may be possible to select for low raffinose and stachyose in strains within a maturity group. Further studies are needed to determine the factors responsible for differences in sugar content. The authors found sucrose and raffinose content positively correlated with oil content and stachyose content positively associated with protein content. Thus, the plant geneticist might have some difficulty breeding out both stachyose and raffinose.

Another antinutritional factor in soybean is trypsin inhibitor. It has been extensively investigated recently since its elimination from soybeans might permit raw soybeans to be used more for

TABLE 47

Oil, Protein, and Selected Sugar Contents of Some Group O-IV Varieties

Variety or line	Oil, %	Protein, %	Total sugar, %	Raffinose, %	Stachyose, %
Amsoy	22.9	37.8	10.8	0.6	2.7
Madison	22.4	39.7	7.4	0.3	1.9
Scott	22.4	36.7	10.1	0.2	2.0
PI 243545	18.6	36.6	10.1	0.5	1.4
Wayne	–	–	9.1	0.9	2.5
Beeson	–	–	9.0	0.7	2.7

feed. Previous work has shown that several inhibitors are present in soybean varieties, but all varieties have trypsin inhibitor. A search of seven wild species of *Glycine* showed that all had inhibitor activity. They were generally very similar in their electrophoretic behavior, except for *G. falcata* which was completely different.[294] Further discussion of trypsin inhibitors can be found in a later section (Nutritional and Physiological Properties).

Aflatoxin in Soybeans?

The report of Shotwell and co-workers[12] that only two samples out of 866 contained aflatoxin and the failure of aflatoxin-producing mold strains to produce much toxin[13] suggests that contaminating grains or weeds may be responsible for such toxins in soybeans. Bean et al.[295] found, however, that 12 out of 24 samples that showed mold growth in the field gave fluorescent materials that tested positively for aflatoxin. Other workers failed to confirm the presence of aflatoxin in Bean's samples. Thus, although soybeans may contain aflatoxin for one reason or another, the possibilities appear to be very low.

EDIBLE OIL PRODUCTS

Deodorization

Recent developments in deodorization have centered around improving the throughput of plants and separation of larger amounts of sterols. The impetus to separating more sterols comes from the demand for stigmasterol and sitosterol for conversion to birth control hormones. Laboratory investigations by a leading processor of deodorizer distillates and shell-drain condensates suggested that optimum conditions for maximum removal of tocopherols and sterols were 275° C, 1-mm pressure of Hg, steam stripping rate of 12% by weight, and a deodorization time of 15 min. Inasmuch as these conditions were not generally delivered by most commercial deodorizers in the late 1960s, the Votator Division of Chemetron undertook pilot investigations in cooperation with General Mills (processor of distillate), Durkee Foods, Division of SCM (processor of edible oils), and Elliot Co., a Division of Carrier Corp. Votator had seen the need for an improved jet steam distributor that would permit higher rates of steam throughput without excess oil entrainment and loss[38] and had found an improved design[296] for a new deodorizer tray that replaces two or three existing trays in the Bailey[35] type deodorizer. The cooperative studies showed that rapid removal of sterol was feasible with maintenance of oil quality. Data from a conventional tray as well as two runs with the new tray at different temperatures and steaming rates are shown in Table 48. As a result of this successful pilot investigation, several new or modified deodorizers with four trays instead of the usual six or seven have now been installed and are operating.

In a development related to the importation of large amounts of palm oil, high-temperature deodorization processing of palm oil is now being carried out commercially without alkali or acetic anhydride refining. Although not practiced on soybean oil, the antipollution feature of the system, as well as the increased yields of oil and high-grade distilled fatty acids, may interest oil refiners. The essential steps are: (1) acidulation to complex trace metals; (2) vacuum bleaching and filtration to remove precipitates (trace metal complexes); and (3) high-pressure steam deodorization at 275°C at 4-to 5-mm pressure to remove fatty acids and improve flavor.[297]

TABLE 48

High Temperature Deodorization of Vegetable Oil, Mostly Soybean Oil

Example	Run I	Run II	Bailey type conventional
Production rate, lb/hr	10,190	11,950	7,500
Charge, lb	3,200	3,750	3,750
Deodorization time, min	19	19	60
Deodorization temp., °C	275	275	240
Pressure, mm-Hg	5.7	1.8	7
Distillate recovery, %	0.724	0.65	0.174
Stripping steam, lb/hr	315	250	225

TABLE 49

Effectiveness of TBHQ in Cottonseed (CSO) and Soybean Oils (SBO) and Potato Chips Fried Therein

Oil	AOM stability, hours to PV 70	Oven test at 145°F (62.8°C) on potato chips, hours to rancidity
Untreated CSO	8	11
CSO + 0.02% TBHQ	34	23
Untreated SBO	11	8
SBO + 0.02% TBHQ	42	25

An Antioxidant for Soybean Oil

For many years it has been known that hydroquinone and its derivatives were superior antioxidants for polyunsaturated oils, such as soybean oil. A search for a sufficiently nontoxic derivative for use as an antioxidant was carried out by Eastman Chemical Products, Inc., a company manufacturing and marketing chemicals for the food and related industries. Their work culminated in the issuance of regulations by the Food and Drug Administration that permit the use of mono-tertiary-butylhydroquinone (TBHQ)[298] or its combinations with other antioxidants up to 0.02% based on the weight of the fat, oil, or fatty content of a food product.[299] This antioxidant at the 0.02% level quadruples the stability of soybean oil, as measured by the AOM Stability test,[300] in terms of the hours to reach a peroxide value of 70 milliequivalents of peroxide per kilogram of fatty material. It appears superior to propyl gallate in increasing the hours of stability in the AOM test. Also, the oven storage life at 115°F (62.8°C) of potato chips fried in soybean oil containing TBHQ is tripled. Table 49 reports selected data on cottonseed and soybean oil.[301]

Flavor Stability of Soybean Oil

Copper oxide catalysts have been shown to give improved edible oils for salads and frying.[63] The use of copper oxide-chromium oxide catalyst in preliminary plant trials was carried out successfully by a large refiner of soybean oil.[302] Problems encountered in these trials were: (1) activity of the catalyst was lower than nickel, and longer hydrogenation times were required than with nickel catalyst; (2) treatments to remove copper contamination of the oil were needed so that the oil was sufficiently stable for both salad and cooking oil use; (3) care in handling of catalysts would be highly desirable since contamination of the copper catalyst with a small amount of nickel makes it behave like a nickel catalyst,[303] and copper contamination of nickel catalyst could lead to copper contamination of nickel-reduced oils, and (4) winterization of the oil was still required although yield of liquid oil was much higher. Private reports indicate that a new plant is now undergoing large-scale runs with copper catalysts.

In related studies it was found that both nickel and copper cause considerable movement of unsaturation under conditions for efficient removal of linolenate ester.[304] However, more conjugatable diene (mostly linoleic acid) was retained in the copper-reduced oil than in the nickel-reduced oil. Data are shown in Tables 50 and 51. On oxidation the isomeric acids probably are precursors of new flavorful aldehydes and ketones. Additional information on double bond movement is reported by Vigernon and co-workers.[305] Procedures for reducing hydrogenation times and amounts of copper catalyst as well as removal of trace amounts of copper by double bleaching and citric acid treatment have been described.[306,307]

Oil from Field-damaged Beans

For many years it has been known that the green color of frost-damaged beans could be substantially removed by the use of acid-activated clay containing about 10% activated carbon. Field damage increases free fatty acids in the crude oil[308] and lowers flavor quality of the refined oil.[309] During the 1971 growing season farmers in North Carolina experienced one of the worst harvest seasons on record, including heavy rains early in November.[310] Edible soybean oil of poor quality resulted whenever the heavily damaged beans were processed. One outlet for such material is alkyd and other resins for inedible use. Stability of the oil was so poor that Evans et al.[311] undertook chemical studies to learn more about it. Apparently field damage not only leads to higher free fatty acid content in the oil but also to abnormally high iron (a pro-oxidant) content. Some of this iron results from contact of the fatty acids with the iron in the bean, but much of it probably comes from contact with iron equipment. Table 52 cites data obtained by Evans and co-workers.

TABLE 50

Fatty Acid Composition of Copper- and Nickel-Reduced Soybean Oils

Acid	Copper	Nickel
C18:0	4	5
C18:1	42	43
C18:2	43	39
C18:3	0.3	3
Calc. IV	111	111
% trans	15	12
Conjugatable diene	41	33
Pressure at 170°C, psi	30	10
Catalyst, %	1	0.03

TABLE 51

Location of Double Bonds in Monoene Isomers* in Copper- and Nickel-Reduced Soybean Oils

Position of double bond	Copper-reduced	Nickel-reduced
Δ7	0.3	0.3
Δ8	2.1	2.1
Δ9	22.1	26.0
Δ12	2.6	5.3
Δ13	1.6	0.8
Δ14	0.9	0.3
Δ15	0.6	0.4
Iodine value (calc.)	113	111
Pressure, psi	30	10
Catalyst, %	3	0.03

*Other monoene isomers present, such as Δ17.

A chromatographic refining loss test run on crude oils from "normal," "abnormal" (assumed to be from damaged beans), and "damaged" beans showed that 46 to 93% of the iron was removed from "normal" and "damaged" oils whereas only 7 to 30% of the iron was removed from "abnormal" degummed oil. Refining with 0.1% excess alkali and with 0.25% acetic anhydride removed 98 and 94% of the iron, respectively, from oil from damaged beans containing abnormally high iron contents. Data also indicated a direct relationship between free fatty acid and iron contents with correlations of 0.74 and 0.88, but a very low correlation between phosphorous and iron contents.

A crude oil (5.7% free fatty acid) commercially extracted from damaged beans was subjected to various laboratory procedures of degumming

TABLE 52

Analysis of Soybean Oil from Beans in Normal and Damaged Conditions[311]

	No. of samples	Phosphorous, ppm	Iron, ppm	Copper, ppm
Crude				
Normal	6	5.70–800	0.9–2.9	0.03–0.05
Abnormal and damaged	11	312–910	2.6–6.1	0.03–0.08
Refined				
Normal	4	1.7–15	0.06–0.2	–
Damaged	1	5	0.1	–
Deodorized				
Normal	4	0.6–5.3	0.08–0.2	0.01
Damaged	2	4.6–59	0.1–0.7	

combined with alkali refining at 0.1 and 0.5% excess alkali, vacuum bleaching with 0.5% activated earth, and deodorization at 210°C for 3 hr. None of these combinations gave an acceptable salad oil. Flavor scores were 4.7 to 5.7. By increasing the deodorization temperature to 260°C, significant increases in flavor scores were achieved. Hydrogenation-winterization and carbon bleaching combined with deodorization at 260°C also improved initial flavor scores.[312] Oxidative stability was poor, but addition of antioxidants was not tested.

Flavor Components in Soybean Oil

The list of compounds known to result when edible soybean oil is oxidized is a very long one. Smouse and Chang[64] reported over 70 volatile acidic and nonacidic compounds. Meijboom and Stroink[313] of Unilever have added a new one that merits serious consideration as a prime candidate for the fishy off-flavor of all oils containing linolenic acid or its ω-3,6,9 structure. 2-*trans*, 4-*cis*, 7-*cis*-Decatrienal was isolated as its 2,4-dinitrophenyl hydrazone and regenerated to produce the fishy off-flavor. It was compared directly with the flavor of a synthetic sample of the trienal. This report helped to clarify earlier ones from Badings of the Netherland Institute for Dairy Research regarding both this trienal and 4-*cis*-heptenal.[314] The trienal is believed by both groups to play a leading part in the so-called "trainy" or fishy off-flavor. 4-*cis*-Heptenal at 1.5 µg/kg may very well be cream-like,[315] but contribute to off-flavor of soybean[65] and other fatty materials at higher concentrations of 15 µg/kg.

CONVERSION TO EDIBLE PROTEIN PRODUCTS

Direct conversion of soybean proteins into foods in the United States is still very small as compared to usage of soybean oil. Present utilization of soybean proteins in foods is equivalent to only about 1% of the crop. This segment of the industry, however, has a large potential for growth and has expanded in the last four years.

Production and Producers

Production estimates of the principal soybean proteins for 1973 and recent prices are given in Table 53. As of May 1974 some processed products were being sold on an allocation basis, but the established processors are expanding their production facilities and several new concerns have entered the field. Table 54 lists the principal U.S.

TABLE 53

Selling Prices and Production Estimates for Edible Soybean Protein Products

Protein product	Price per pound,* cents	Production for 1973[†] million pounds
Flours and grits	11–14	450–600[‡]
Concentrates	30–38	55
Isolates	58–64	50
Textured products		
Extruded flours	21–24	} 110
Spun isolates	50 and up	

*Prices as of October 1974.
[†]Estimates by Lockmiller.[316]
[‡]Includes edible grade used for pet foods.

TABLE 54

Principal U.S. Producers of Soybean Protein Products*

Producer	Grits and flours			Concentrates	Isolates	Textured products	
	Defatted	Low-fat**	Full-fat			Flours	Isolates
Anderson Clayton Foods					+		
Archer-Daniels-Midland Co.	+	+	+	+†		+	
Cargill, Inc.	+	+‡				+	
Carnation Co.					+		
Central Soya Co.	+	+	+	+	+	+	
Far-Mar-Co	+			+		+	
General Mills, Inc.						+	+
Grain Processing Corp.				+	+		
Griffith Laboratories				+		+	
Lauhoff Grain Co.	+					+	
Miles Laboratories, Inc.						+	+
Nabisco						+	
National Protein Corp.	+					+	
Ralston Purina Co.					+	+	+
A. E. Staley Mfg. Co.	+	+		+†	+♯	+	
Swift & Co.	+			+		+	

*Updated from Lockmiller.[317]
**All producers of low-fat flours also offer lecithinated flours.
†Plant under construction.
‡Offers lecithinated but not low-fat flour.
♯Offers mainly pepsin-modified isolates.

producers and types of protein that they supply. The most rapidly growing segment of the business has been the textured protein products.

World-wide demand for protein has caused shortages of traditional animal proteins and dramatic increases in prices. This situation, in turn, has resulted in the development of soy-based protein alternates. For example, milk replacements for use in baked goods, beverages, ice cream, and confections are now available.

New Processes

Widespread interest in soy proteins in the past decade has stimulated development of a variety of new processes, many of which have been described in the last five years. Selected examples are reviewed here.

Full-fat Products

University of Illinois workers have concentrated on development of foods in which the entire soybean, including the seedcoat, is used.[318,319] They recommend blanching soybeans by boiling to inactivate lipoxygenase before disrupting or damaging the cellular structure of the seed. If lipoxygenase is allowed to remain active and the seed tissue is disrupted, enzyme action is believed to result in the generation of painty or beany flavors. Prototype foods containing whole soybeans, such as canned soybeans and chicken, have been prepared. Alternatively, the blanched beans can be macerated and drum-dried to obtain a flake-like product.

Two processes have been described for converting full-fat soy flours into beverage bases. In the first,[320] extrusion-cooked flour is dispersed in water, blended with soybean oil and emulsifier, colloid milled, homogenized, and spray-dried. The resulting powder is then blended with sugar, salt, flavor, minerals, and vitamins to give the beverage base. In the second process[321] raw full-fat flour is slurried in dilute sulfuric acid at a final pH of 3.5 and then heated quickly with steam to inactivate lipoxygenase. After cooling, the lipid-protein-polysaccharide curd is washed with water to remove the soluble sugars and adjusted to pH 9 where another quick cook is given to inactivate any trypsin inhibitor that survived the first cooking at low pH. After neutralizing, the lipid-protein complex is colloid milled, homogenized, and centrifuged to yield the beverage base.

Defatted Flakes and Related Products

Although hexane is widely used to extract oil from soybeans, it does not remove all lipids.[322] Increasing attention is being given to solvents that remove the residual lipids not extracted by hexane. Extraction of hexane defatted flakes with hexane:alcohol mixtures removes additional lipids and simultaneously removes beany and bitter flavors characteristic of raw defatted flakes.[268,323] Hexane:ethanol in a ratio of 82:18 (v/v) is an effective solvent and has the advantage of causing a minimum of protein denaturation. Flakes reextracted with hexane:ethanol can also be used to prepare protein isolates with improved flavor.

Another solvent extraction procedure reported to be effective in removing residual lipids is ethanol followed by a 1:1 (v/v) mixture of ethanol:chloroform.[324] When the extraction was applied to full-fat flour, alcohol removed 20.6% of solids and ethanol:chloroform removed an additional 5.6% of solids. The latter is believed to contain bound lipids that are bitter and have a mouth-coating effect when tasted. A bland meal is claimed for this extraction procedure, but inordinately long extraction times appear to be needed. Alcohol extraction is carried out for 2 hr and the alcohol:chloroform extraction takes about 22 hr.

Concentrates

The conventional acid-washing process for making protein concentrates[138] has been modified by adding a plant gum such as carrageenan during the acid-washing step.[325] The plant gums form insoluble complexes with the acid-soluble (whey) proteins;[326] consequently, these proteins are not lost during the acid-washing step.

A two-phase solvent system of hexane, methanol, and water in a ratio of 10:7:3 has been used to extract full-fat soybean flakes to simultaneously remove the oil and oligosaccharides plus minor nonprotein constituents.[327] The resulting product contained 65 to 73% protein and is a protein concentrate similar to that obtained by aqueous alcohol extraction of defatted flakes.[135,136] Presumably, the hexane and methanol remove residual lipids as observed when defatted flakes are extracted with this two-component solvent.[268,323]

Another process that utilizes hexane:alcohol as a solvent for residual lipids is described by Hayes and Simms.[328] Full-fat flakes are extracted with hexane, and ethanol is then added to the flakes saturated with hexane; the usual desolventization step is omitted. The hexane:ethanol mixture removes the residual lipids, and the drained flakes are then desolventized to remove the hexane selectively. Next, additional ethanol and water are added to the alcohol-wet flakes to extract the soluble sugars and minor constituents. The flakes are then desolventized to yield a protein concentrate.

Ethanol:water:hexane in the ratio of 75:15:10 has been used to extract defatted flakes to yield a product that approaches a protein concentrate in composition. The extracted flakes are claimed to be tasteless and contain 65% protein.[329]

Hoer and Calvert[330] have also modified the isoelectric wash process[138] for making concentrates. Defatted soybean flakes are slurried in water to dissolve the proteins and other soluble constituents. The slurry is then adjusted to pH 4.5 to form a curd of isoelectric protein and water-insoluble polysaccharides. After centrifuging and thorough washing, the solids are adjusted to pH 6.8 and then passed to a jet cooker where the temperature is raised to 310° F and held there for about 5 sec. Injection into a vacuum chamber flashes off volatile undesirable flavor components, and after spray drying a protein concentrate is obtained. Precipitation of the proteins on the surface of the insoluble polysaccharides is claimed to be a novel feature of the process, but it is likely that the proteins not denatured by isoelectric precipitation redissolve when the pH is readjusted to 6.8 before the jet cooking step. Heat denaturation during jet cooking is probably the most important factor in altering functional properties of the resulting protein concentrate as compared to the conventionally processed concentrate made by isoelectric washing:

Another variation[331] of the isoelectric washing process for preparing protein concentrates consists of stirring defatted flakes in water for 15 to 30 min and then adjusting the pH to 4 to 5. (This step is identical to that proposed by Hoer and Calvert.)[330] The isoelectric slurry is then passed through a shearing-type comminutor (MG Model Urschel Comitrol with Microcut-Head attach-

added to pork sausage with little effect on texture, but separation of water from the sausage was reduced.

PROPERTIES OF SOY PROTEINS

Research on various properties of soy proteins of importance for food uses is continuing in many laboratories. Functional and nutritional properties plus the flavor problem have received major attention.

Functional Properties

Several reviews of this subject have appeared. Briskey[345] discussed methods for measuring functional properties of proteins, some of which may be applicable to soy proteins. Hermansson[346] reviewed her studies on protein-water interactions with emphasis on protein solubility and swelling during sorption of water.

Solubility

Hermansson[346] measured solubilities of a commercial soybean protein isolate, sodium caseinate, milk whey protein concentrate, and fish protein concentrate as a function of pH in 0.2 M sodium chloride. The soy isolate showed the expected minimum in solubility in the pH 4 to 5 region but was only 25 to 30% soluble in the pH range of 6.5 to 8.0. High solubility did not occur until the pH was raised to above 11 which is well outside of the useful range for foods. The effect of ionic strength at pH 7 was also reported. Solubility in water was about 55% and dropped to around 28% when salt was added to 0.2 M. Increasing ionic strengths caused a slow rise in solubility followed by a slow decline. Solubilities of sodium caseinate and whey protein concentrate were fairly insensitive to changes in ionic strength up to 2 M salt.

Solubility of soybean globulins as a function of pH and ionic strength is reported by van Megen.[347] Below critical values of ionic strength, he observed an unusual liquid-liquid phase separation. This two-phase system consisted of a protein-poor upper layer (supernatant) and a viscous protein-rich lower layer. Above the critical ionic strength, the system was homogeneous. Tombs[339] refers to the two layers as mesophases and uses them to spin isolates into fibers as described earlier.

Hagenmaier[348] has reported limited data for water solubilities of two commercial soybean protein isolates. One isolate had a solubility of 22% and the other 40% at pH 6.0.

Water Absorption and Swelling

Hermansson[349] measured swelling properties of soy isolate by following the water uptake of the protein in an apparatus described by Baumann.[350] The Baumann apparatus consists of a water-filled funnel connected to a horizontal capillary tube which is calibrated so that water uptake can be followed. The funnel also contains a fritted glass disc that is in contact with the water. In the test the dry protein is sprinkled on a wetted filter paper placed over the glass disc and the water uptake is followed as a function of time. The soy isolate swelled more than sodium caseinate or whey protein concentrate. Swelling of heat-treated isolates showed some correlation with textural properties of meatballs containing these isolates. Hermansson[346] also measured the sorption of water vapor by a soy protein isolate at three water activities. Although the swelling properties of the isolate increased significantly when the protein was heated, water sorption increased very little.

Hagenmaier[348] determined water sorption of soy protein isolates at 84% relative humidity and compared them with other plant proteins plus several animal proteins. Plant proteins tended to bind less water than animal proteins, and water binding appears to be related to the content of polar side chains — hydroxyl, amino, and free carboxyl groups.

Fleming et al.[351] measured water absorption of sunflower and soybean flours, concentrates, and isolates by making 10% slurries in water or 5% sodium chloride and then centrifuging. The decrease in volume of free liquid was taken as the amount of water absorbed. Of the soybean products, isolates had the highest water absorptions followed by concentrates which in turn had higher absorptions than the flours. This test, however, has serious limitations. For example, an isolate that is completely soluble will show no apparent water absorption, yet when it is incorporated into a food product that is heated it may be insolubilized or gelled and hold significant amounts of water.

Smith et al.[352] studied effectiveness of various protein additives, including six soy proteins, as emulsion stabilizers in frankfurters. Water-holding capacity was measured under conditions of salt concentration, pH, and heat treatment that simulated frankfurter manufacturing. Nonetheless, the water-holding capacity measurements were of little value in predicting performance characteristics of the protein additives in frankfurters.

Viscosity

Data for soy protein isolate in water and 10% protein gels formed by heating are reported by Hermansson.[349] The results are in general agreement with earlier data reported by Circle et al.[100] Isolate heated in sodium chloride formed much weaker gels than when heated in water.

Fleming and co-workers[351] compared viscosities of sunflower and soybean flours, concentrates, and isolates slurried in water and 5% sodium chloride solutions. Protein concentrations from 5 to 20% were studied. The protein samples were also given a "pH activation" treatment consisting of adjusting the pH to 12.2, holding for a few minutes, and then lowering the pH to 6.0. Generally, pH activation caused an increase in viscosity. Presumably, the high pH causes dissociation of the globular proteins into subunits and unfolding of the native structures into more disoriented entities.

Emulsification

Crenwelge et al.[353] measured emulsion capacities of soybean and cottonseed protein concentrates in a model system of water, protein, and corn oil. Emulsification was conducted in a Waring Blendor® by adding oil until emulsion inversion occurred as measured by a sharp drop in current required by the blender motor. Variables such as blender speed, rate of oil addition, pH, and protein concentration were optimized, but no attempt was made to relate the model to a food system.

Smith and co-workers[352] compared emulsifying capacity and emulsion stability of protein additives in model systems and frankfurter formulations. Careful study of their results raises serious questions about the usefulness of model tests for evaluating effectiveness of protein additives in this type of application. Their work also indicates that protein solubility of the additive is not always required; fish protein concentrate in which the protein was completely insoluble gave stable emulsions. Apparently, insoluble particles of sufficiently small size can collect at the oil-water interface and prevent coalescence of fat globules.

Film Formation

When soy milk is heated in flat shallow pans at about 90°C a cream-yellow surface film forms which can be lifted off with glass rods and hung up to dry. The dried product called yuba or soy milk skin is a traditional food made in the Orient. Wu and Bates[354-356] have made an extensive study of this film-forming process. Model system studies showed that protein is essential for film formation. Typical films made from soy milk contained about 55% protein, 26% neutral lipids, 2% phospholipids, 12% carbohydrate, and 2% ash on a dry basis. Approximately 70% of the total soybean protein is recoverable as films under optimum conditions. Protein-lipid films were also obtained from "milks" prepared from peanuts and cottonseed, whereas cottage cheese whey with or without added lipid formed little or no film on heating. The film-forming proteins of cow's milk apparently are removed during cheese making.

Texture

Patents describing new methods for producing textured products were discussed earlier (New Processes); hence, only basic studies on textural properties will be reviewed here.

Extrusion of defatted soybean meal has been studied over the temperature range of 110 to 190°C by Cumming et al.[357] Effects of processing temperature on product density, extent of rehydration, shear force, work of shearing, and breaking strength were examined. Scanning electron microscopy showed that extrusion converts soybean meal into a spongy, fibrous mass. Changes in solubility and electrophoretic mobility of the proteins in soybean meal as a result of extrusion are also reported by Cumming and co-workers.[358]

Textural properties of spun protein isolates have likewise been investigated and compared with corresponding properties of meat.[359] Dry fibers were rehydrated in boiling water and measurements were made of load elongation, breaking strength, break elongation, elasticity, and stress relaxation. Breaking strength and break elongation of the soy fibers were considerably higher than similar properties of cooked meat.

When spun soybean protein fibers are stored, they become tough, inextensible, and turn from a cream-white to a gray color. The content of sulfhydryl groups decreases on storage, and formation of disulfide crosslinks is believed to cause the observed changes in physical properties.[360]

Previously noted differences in the texture of tofu prepared from crude 7S and 11S proteins[111]

ment), centrifuged, and washed to remove the soluble sugars. The washed material can be dried or neutralized, heated, and then spray dried. The shearing treatment is claimed to rupture cell walls and protein body membranes. Although the proteins in some commercially prepared flakes may be difficult to extract, such is not true with carefully prepared flakes where high nitrogen solubility index values suggest a high degree of cell rupture.[332] Moreover, protein bodies in soybeans appear very labile and are difficult to isolate intact even under conditions specifically selected to maintain their integrity.[333]

Isolates

A variety of protein isolates with differing physical properties are now available. Processes used to produce these isolates are proprietary information, but a likely modification is controlled heat treatment. Hoer et al.[334] described modification of sodium soy proteinates by jet cooking and enzyme treatment. Slurries of proteinates are quickly heated (285 to 320° F) in a jet cooker which also subjects the protein to a shearing action. After holding the heated slurry for a short time it is discharged into a lower pressure zone to flash off water vapor plus volatile flavor compounds. Next, the cooled slurry is treated with a proteolytic enzyme for a short time (1 to 30 min). The enzyme is then inactivated by reheating the slurry and the final step is spray drying. The modified isolate is claimed to have high wettability and good dispersibility when mixed with water.

Textured Protein Products

Work on texturization of soy proteins has received considerable attention in the last few years. Gutcho[335] has summarized much of the U. S. patent literature on this subject from 1960 to 1972. Extrusion of defatted flours and spinning of protein isolates are the major techniques used to texturize soy products, but alternative techniques are being sought. Boyer and co-workers[336] mixed isolates or flours with water to form a dough that was then cooked in a microwave oven to cause expansion and development of a cellular structure. By adding appropriate flavors and colors to the expanded materials, they developed fried bacon-like products and extenders for ground meat.

Strommer[337,338] has developed a rotating multichambered device in which a blend of defatted soy flour, protein concentrate, and isolate can be injected with high-pressure steam and then expelled through a nozzle. The expulsion causes puffing and texturization. Sudden release of pressure when the material is expelled through the nozzle apparently causes volatile compounds to flash off; a bland taste is claimed for the textured product. In contrast to ordinary extrusion, this process involves a minimum of mechanical working of the material being treated. Mechanical working occurs only during the brief time that the material passes through the exit nozzle. Residence time of the protein material in the apparatus may be less than 1 sec.

Tombs[339] has described a new variation of the classical spinning process in which the alkaline spinning dope is eliminated. Soy protein isolate is prepared in the usual manner by isoelectric precipitation at pH 4.6 to 4.9, but instead of dissolving the curd with alkali, water and solid salt are added to a final molarity of 0.4 to 0.5. The curd does not dissolve completely, but on centrifuging the soluble portion or mesophase can be separated. The mesophase is then extruded into water at 80° C or higher. Leaching of the salt from the extrudate plus the high temperature coagulates the protein to form the fibers. Eliminating the alkaline spinning dope minimizes the possible formation of lysinoalanine.[340]

New spinning processes for defatted soybean meal[341] and protein isolates[342] have been described. Defatted flakes or protein isolates are made into a thick water slurry in the isoelectric pH region and pumped at high pressure (50 to 5,000 psi) through a heat exchanger. The heat denatured proteinaceous material is then forced through a narrow orifice to obtain small, discrete textured particles or a continuous filament, depending on operating conditions. The process has the advantage of simplicity and eliminates extensive washing steps involved in the conventional spinning process.

Isolates can also be textured by a modification of the ancient Oriental process for manufacturing kori or dry tofu.[343,344] Acid-precipitated soybean protein is dissolved in water at pH 7.0 and heated at 100°C for 1 min. After cooling, calcium chloride is added and the solution is then frozen at −20° C and held at this temperature for 2 hr. Next, the frozen protein solution is stored at −5° C for 1 day, thawed, and washed to yield a spongy product. Up to 40% of the spongy protein was

have also been observed for gels formed by heating dispersions of the two proteins.[361] Gels made from crude 11S protein were firmer and more cohesive than gels made from crude 7S protein. An unusual porous texture results on autoclaving soy protein gels made by heating protein solutions, coagulating with calcium ion, or acid followed by pressing.[362]

Nutritional and Physiological Properties

Reviews of nutritional, biological, and physiological properties of soybeans and soybean proteins by Liener[363] and Rackis[364,365] should be consulted for detailed discussion of these subjects. Items of recent interest include the biological effects of trypsin inhibitors; nutritional properties of soybean proteins in blended foods, textured products, and infant formulas; and the effect of alkali on soybean proteins. Although soy proteins have been fed to experimental animals in countless studies, data are now also accumulating in which human subjects have been used.

Trypsin Inhibitors

Screening of 108 soybean strains and varieties showed a nearly four-fold range in trypsin inhibitor content, but none of the samples were free of inhibitor activity.[366] The Kunitz trypsin inhibitor occurs as a genetic variant that can be detected by polyacrylamide gel electrophoresis.[367] Electrophoretic screening of 1,595 plant introductions in the USDA germplasm collection revealed protein bands corresponding to trypsin inhibitor in all samples. One inhibitor form predominated; it occurred in 1,255 samples[368] and is found in the major commercial varieties presently grown in the United States.[369]

One of the inhibitor variants found in an experimental variety was found to have a lower specific activity than the common variant found in Clark and Harosoy varieties.[370] When the experimental variety was fed raw to chicks, it caused pancreatic hypertrophy plus a lower weight gain than heated defatted meal of commercial origin. The experimental variety, however, gave a better growth response than either raw Clark or Harosoy beans.

When soybean proteins, from which trypsin inhibitors were selectively removed by affinity chromatography, were fed to rats about 60% of the pancreatic hypertrophy and 60% of the growth inhibition noted in proteins containing trypsin inhibitors remained. The residual pancreatic hypertrophy and growth inhibition in the absence of trypsin inhibitors were attributed to resistance of native soybean proteins to proteolytic enzymes.[371]

It is generally advocated that soybean protein products be thoroughly heated in some stage of processing to inactivate trypsin inhibitors, but it is not known what extent of inactivation is needed to make the inhibitors innocuous. An improved assay to measure low trypsin inhibitor activities in heated products has, therefore, been developed to study this problem.[372] The assay was used to determine residual inhibitor in a series of defatted flakes steamed for varying times to obtain graded levels of inhibitor activity.[373] The series of flakes was then fed to rats. No pancreatic hypertrophy was observed with flakes still containing 31 to 45% of the original inhibitor activity; however, maximum body weights, protein efficiency ratios, and nitrogen digestibilities were obtained with samples receiving more heat treatment and having residual inhibitor activites of only 13 to 21%. In contrast to rats and other species, dogs can ingest 15% raw soybean with no apparent deleterious effects on pancreas weights, pancreatic enzyme secretion, or body weights.[374,375]

The importance of trypsin inhibitors in the human diet is still unkown, but residual inhibitor activity appears to be no problem in cooked meat products containing soy protein. Nordal and Fossum[376] found that meat or meat extracts sensitize soybean trypsin inhibitor to heat. A mixture of meat and isolated soybean protein heated for only 5 min at 90°C contained no detectable trypsin inhibitor.

Soybean Proteins in Blended Foods

Several blended foods containing soy flour have been developed and evaluated nutritionally. The best known of these is corn-soy-milk (CSM). Graham et al.[377] fed CSM (68% corn meal, 25% defatted soy flour, and 5% nonfat dry milk) to infants and showed that the protein blend had a biological value about 60 to 65% of that for casein. When they fed "instant sweentened" CSM, which contains a higher ratio of soy protein to corn protein (40% corn meal, 38% full-fat soy flour, 5% nonfat dry milk, and 15% sugar), Graham and co-workers found that the product had an excellent digestibility and nitrogen retention was similar to that for casein.[378] The increase in biological value of "instant sweetened" CSM over CSM was attributed to a more complementary ratio of soy to corn proteins. The modified CSM was recommended as a major source of dietary protein for normal and convalescent malnourished infants and children.

Tests with infants and children indicated that wheat-soy mixtures had satisfactory biological values but poor digestibilities, whereas oat-soy products had an excellent biological value with a digestibility equivalent to that of the wheat-soy products.[379] Macaroni containing 60% corn meal, 30% defatted soy flour, and 10% wheat flour was fed to malnourished children in Brazil for four months. A control group of children received a normal balanced diet containing 25 to 30% milk protein. The corn-soy-wheat noodle had excellent acceptance and supplied 69% of the protein for the experimental group. Both groups recuperated nutritionally, but the control group showed a better response possibly because it had a higher caloric intake than the children on the noodle diet. The macaroni was recommended as a protein supplement.[380]

Soybeans have also been used successfully to enrich tortillas.[381,382] Corn and soybean mixtures in various ratios were cooked in limewater and converted into tortillas in the traditional manner. When fed to rats the soybean-fortified tortillas showed a marked improvement of the protein efficiency ratio. The highest protein efficiency ratio was realized with a 72% corn: 28% soybean mixture. Organoleptic evaluations indicated that the presence of soybeans in tortillas is not significantly detectable at soybean levels below 20%.

Nutritional Value of Textured Soybean Proteins

An extruded soy flour with and without 1% DL-methionine was compared with beef in nitrogen balance studies with adult men.[383] At an 8.0-g nitrogen intake per day all three diets maintained the subjects in positive balance, whereas at a 4.0-g nitrogen intake per day, subjects went into negative nitrogen balance. At the lower nitrogen intake beef, textured soy flour, and the same soy flour fortified with methionine had respective nitrogen balances of -0.30, -0.70, and -0.45 g, N per day. In a second study mixtures of beef and textured soy flour were fed at varying ratios at an intake level of 4.0 g of nitrogen per day.[384] Increasing the level of textured soy flour led to a progressively more negative nitrogen balance; this result confirms that beef protein is nutritionally superior to soy flour proteins. No mutual supplementation was observed, however, and methionine was also suspected to be limiting in beef. The comparison was somewhat biased in favor of the beef because the nitrogen-to-protein conversion factor for beef is about 6.25, whereas the factor for soy proteins is about 5.7. Differences in nonprotein nitrogen content between beef and soy flour also need to be considered.

Ingestion of 20 g of protein derived from textured soy flour in place of meat protein in a normal diet had no significant effect on nitrogen balance in human subjects, although a higher positive nitrogen balance was obtained on the meat-containing diet.[385] Soy protein fibers were compared with casein-lactalbumin in which both diets were adjusted to the provisional FAO essential amino acid pattern by adding crystalline amino acids.[386] Nitrogen balances in young men showed no significant differences; both protein sources appeared to be equally well-utilized.

Soybean Proteins in Infant Formulas

Isolated soy protein fortified with DL-methionine was fed in initial therapy of malnourished infants and children and during convalescence. Growth rates, as well as absorption and retention of nitrogen, were equivalent to those from modified cow's milk.[387]

In another study 13 normal female infants from 8 through 11 days of age were placed on a diet in which soy protein isolate fortified with DL-methionine provided 90% or more of the dietary amino acids.[388] Clinical observations, serum albumin concentrations, and growth rates for the infants on the soybean protein diet were similar to those of infants fed milk-based formulas that provided greater intakes of protein.

Effect of Alkali Treatment on Soy Protein

De Groot and Slump[340] treated soybean meal with alkali at pH 12.2 for 4 hr at 40°C and found a decrease in lysine content and the appearance of lysinoalanine. Net protein utilization of the alkali-treated soy protein decreased and 40% of the ingested lysinoalanine was excreted mainly in the feces. Isolated soy protein was also treated with alkali at various pH values, times, and temperatures and then analyzed for lysinoalanine. As little as 1 hr at pH 12.2 and 40°C resulted in formation of lysinoalanine, whereas 4 hr as pH 8 and 40°C did not. Detectable amounts of lysinoalanine were noted after 4 hr at pH 10 and 40°C. Examination of tissues in rats fed alkali-treated protein revealed no changes except in the kidneys. A few of the female rats exhibited nephrocalcinosis which was overcome by adding supplemental calcium to the diet.

Woodard and Short[389] recently confirmed and extended this work. Earlier studies by Woodard showed that feeding industrial grade soy protein isolate (modified by alkali treatment to alter viscosity and adhesive properties) to rats caused cytomegalia in the straight portion of the proximal tubule of the kidney. Edible grade soy protein showed no effect on the kidney cells, but after treatment with 0.1 N sodium hydroxide at 60°C for 8 hr it gave a response similar to that of the industrial grade of protein. Amino acid analysis of the industrial and modified edible proteins revealed the presence of lysinoalanine in both samples; however, the crucial experiment of feeding authentic lysinoalanine was not tested. Further study is needed, but the results thus far emphasize the need to carefully control alkali treatment in processing of soy proteins for edible purposes.

Flavor Studies on Soy Products

Flavor continues to be a problem in the use of soy proteins in foods, particularly in bland products where flavor of ingredients is readily detected. Nonetheless, progress is being made and soy proteins are now found in a wide range of foods. For example, liquid coffee whiteners containing soy protein isolate are available, and a major company that has traditionally used milk proteins recently began adding soy proteins to some of its products and is also offering a soy protein isolate for sale. Several reviews have described current status of the soy flavor problem.[390-392]

Organoleptic Evaluation of Commercial Protein Products

Taste panel evaluation of odors and flavors of commercial soy flours, protein concentrates, and isolates available in 1970 are summarized in Table 55.[253] Samples were evaluated as 2% dispersions in water and scored on a scale of 1 to 10 where 1 is a strong odor or flavor and 10 is bland. A raw hexane-defatted flour prepared in the laboratory was included for comparative purposes; the raw

TABLE 55

Odor and Flavor Scores for Soybean Protein Products

Samples	Odor scores*	Flavor scores*
Defatted flours	5.8–7.5	4.1–6.7
Concentrates	6.4–7.4	5.6–7.0
Isolates	6.8–7.7	5.9–6.4

*Scoring is 1 to 10 where 1 is strong and 10 is bland.
(From Kalbrener et al.[253])

flour had the lowest odor and flavor scores. An inverse relationship between flavor score and protein solubility was obtained for the flours. Commercially, moist heat treatment is used to improve the flavor characteristics of these products, but the proteins are insolubilized in the process. The highest flavor score was given to one of the concentrates. Flavor descriptions included beany, bitter, nutty, and toasted. Flavor thresholds for selected products were also determined. Respective thresholds for beany and bitter flavors in the raw flour were 0.03 and 0.04% as compared to 1.25 and 3.0% for one of the isolates. The compounds responsible for beaniness and bitterness in raw defatted flours obviously are very potent and difficult to remove.

Origin of Flavor Compounds

Several lines of evidence point to the lipids as a source of undesirable flavors, and additional work implicating lipoxygenase in the formation of flavor compounds has been reported. Nelson et al.[318] claim that intact soybeans are free of undesirable flavors and the flavor compounds form through lipoxygenase-catalyzed oxidation of lipids as soon as the cellular structure is disrupted. They recommend blanching the beans before crushing them to prevent formation of the unwanted flavor constituents.

Incubation of linoleic acid with partially purified soybean lipoxygenase in the presence of oxygen resulted in the formation of 3.4% monocarbonyl compounds in the following amounts:[393]

Aldehyde	Mole %
n-Pentanal	5
n-Hexanal	45
n-Hept-2-enal	10
n-Oct-2-enal	5
n-Nona-2,4-dienal	5
n-Deca-2,4-dienal	26
Unidentified	4

As noted earlier in identification of volatile compounds from soy milk,[264] hexanal was the major aldehyde isolated.

Involvement of hydroperoxides of linoleic and linolenic acids in the origin of some of the flavors of soybean products was indicated by tasting dilute dispersions of the purified hydroperoxides in water.[394] Linoleic acid hydroperoxide at 50 ppm and linolenic acid hydroperoxide at 10 ppm were scored the same flavor intensity as a 0.25% dispersion of raw defatted soy flour. The flavors detected were assumed to result from decomposition of the hydroperoxides. The major flavor description given was grassy/beany; minor responses given were bitter, astringent, raw vegetable, and fruity. Flavor descriptions for defatted flour did not include the last two terms; hence, the two hydroperoxides do not reproduce the typical soybean flavors exactly, but they do appear to be important precursors for the grassy/beany flavors.

Soybean phosphatidyl choline is a potential source of bitterness in soybean meal products.[395] When the purified phosphatide is autoxidized for two weeks in aqueous dispersion with 1 ppm of Cu^{++}, it develops a bitter taste with a threshold of 0.006%.

FOOD USES OF SOYBEAN PROTEINS

Developmental work on incorporation of soy proteins into new and old food products is continuing, and a number of these items have been introduced in the last few years (Table 56).

Baked Goods

Many of the white breads in the United States now contain 1.5 to 2.0 lb of soy flour/100 lb of wheat flour to replace nonfat dry milk solids which have become almost prohibitively expensive.[396,397] Annually about 50 million lb of defatted soy flour are used in bread, another 14 million lb are added to specialty baked products and crackers, and about 7 million lb of soy flour with different fat contents are utilized in doughnut mixes and cakes.

A variety of nonfat dry milk replacers for baking and margarine applications has been developed recently.[397-399] Several of them consist of cheese whey-soy flour blends; some also contain soy protein isolates in combination with milk proteins.

Details for incorporating up to 24% full-fat soy flour into bread have been described.[400] By adding 0.5% sodium stearoyl-2-lactylate the adverse effects of soy flour on loaf volume and grain score can be overcome. The process does not appear to be used commercially, nor have organoleptic evaluations of such fortified breads been reported.

Soy flours (defatted or full-fat) and protein isolates can also be used to raise the protein content and improve the balance of the amino acids in cookies.[401] Surfactants such as sodium stearoyl-2-lactylate are important for successful fortification of cookies with soy proteins.

Meat Products and Analogs

Two factors have contributed to substantial increases in U.S. consumption of soy proteins in meat products: (a) use of textured soy proteins in the National School Lunch Program beginning in 1971 and (b) introduction of ground beef-soy blends in supermarkets in March 1973. Both developments resulted from the increased costs of animal proteins, particularly meats.

The National School Lunch Program consumed about 9 million lb of textured soy proteins in the 1971 to 1972 school year, and the amount doubled in the following year.[402] These products are also used in other institutional feeding programs. Marketing of ground beef-soy blends in 1973 in a cooperative effort by General Mills Company and Red Owl Stores, Inc. in Minneapolis, was the first major presentation of textured soy products at the retail level. Initial consumer acceptance was excellent, and by the end of 1973 the blends had penetrated 50% of the nation's retail markets.[403] Since then ground beef prices have dropped and sales of beef-soy blends have also decreased.

A number of convenience food items, such as casserole mixes containing textured soy proteins, have likewise been marketed. About 110 million lb of textured products were produced in 1973 (Table 53) and predictions are that the market will double each year during the rest of the 1970's.[404]

In 1974 Miles Laboratories introduced four meat analogs in the retail markets. The analogs are sold in frozen form and contain spun soy isolate fibers along with other protein sources such as wheat, yeast, oats, and egg albumin.[405] Miles

TABLE 56

New Commercial Food Products Containing Soy Proteins

Company	Protein form used	Product
Carnation Co.	Isolate	Instant Breakfast (chocolate flavor)
	Isolate	Hot Cocoa Mix
Colgate-Palmolive Co.	Soybeans	Bambeanos® (roasted soybean snack)
Colonial Baking Co.	Defatted flour	New World Bread (special formula bread with granola added)
Creamette Co.	Defatted flour, acid hydrolyzed protein	Protein-ettes® (textured vegetable protein-simulated meat for casseroles)
General Mills, Inc.	Concentrate and isolate	Breakfast Squares® (high-protein breakfast bar)
	Defatted flour	Chipos® (barbecue-flavored potato chips)
Gooch Foods, Inc.	Defatted flour	Red Skillet TVP® Dinner (one-pan meal mix)
Interstate Brands Corp.	Defatted flour	Butternut Bread with Buckwheat (special formula bread)
Kroger Co.	Defatted flour	Burger·Pro® (ground beef-textured soy flour blend)
Miles Laboratories	Isolate	Breakfast Patties (simulated pork sausage patties)
	Isolate	Breakfast Links (simulated pork sausage links)
	Isolate	Breakfast Slices (simulated ham-Canadian bacon)
	Isolate	Breakfast Strips (simulated sliced bacon)
The Pillsbury Co.	Defatted flour, isolate	Bundt® Chocolate Macaroon (cake mix)
	Isolate	Food Sticks (chocolate flavored snack food)
A. E. Staley Mfg. Co.	Defatted Flour	Burger Bonus® (textured soy flour for extending meat)
Standard Brands	Soybeans	Planters® Soy Nuts (roasted soybean snack)

Laboratories also markets meat extenders that contain spun isolate which are capable of being used at levels up to 45% as compared to textured soy flours which are recommended to be used only at levels of 25%.[403] Meat analogs were reviewed in detail recently.[406]

Instant Breakfast Items

A variety of convenience items designed for a quick morning meal or in-between-meal snack have appeared on the market. A chocolate-flavored product to be mixed with milk contains soy protein isolate while another in the form of semimoist bars contains protein concentrate and isolate. Two problems limiting use of soy proteins in these products, particularly beverage items, are incompatibility of functional properties and residual soy flavors.[407]

Snack Foods

Roasted soybeans are now available from several companies. Protein-fortified snack items anticipated several years ago have not materialized to any significant extent. Several were test-marketed but withdrawn again. A fabricated potato chip containing soy flour was introduced recently, but the soy flour is probably a functional ingredient rather than a significant protein source.

Legal and Regulatory Aspects

This situation was reviewed for the United States,[408-410] United Kingdom,[411] and Japan[412] at the World Soy Protein Conference held in Munich in 1973. Problems and philosophy of regulating new protein foods were also discussed recently.[413]

In the United States the Food and Drug Administration is developing definitions for soybean protein products and regulations for control of their use in foods.[408] This is not an easy task because a wide variety of protein preparations is already available, and many more can be anticipated in the future as research and development move forward. Many of the present products are only prototypes of future foods.

Closely intertwined with regulating the addition of soy proteins to foods is the need for reliable methods for measuring the amount of soy protein added. Many methods for determining soy proteins in meats have been proposed, but most are of limited usefulness because they are sensitive to the processing conditions employed in the meat industry.[414] The U.S. Department of Agriculture presently determines soy flours and protein concentrates in meat products by measuring the amount of hemicellulose present, whereas isolates are monitored by analyzing for titanium which is added to isolates during manufacture as titanium dioxide at a level of 0.1%. A recent study shows that defatted soy flour is high in magnesium content whereas beef is comparatively low. Determination of magnesium was proposed as a method for analyzing beef-soy flour blends.[415] Many electrophoretic methods have also been proposed; one just published appears promising.[416]

Present methods are limited to a single soy product type; they cannot distinguish between mixtures of flours, concentrates, and isolates. One textured product currently available contains all three protein types and another contains protein concentrate, spun isolate, plus isolate.

REFERENCES

1. Williams, L. F., in *Soybeans and Soybean Products*, Vol. 1, Markley, K., Ed., Interscience, N.Y., 1950, chap. 3.
2. Saio, K. and Watanabe, T., *J. Food Sci. Technol.* (Japan), 15, 290, 1968.
3. Tombs, M. P., *Plant Physiol.*, 42, 797, 1967.
4. Wolf, W.J., *J. Am. Oil Chem. Soc.*, 47, 107, 1970.
5. Pfeifer, V. F., Stringfellow, A. C., and Griffin, E., L., *Am. Miller Process.* 88(8), 11, 1960.
6. Kawamura, S., *Tech. Bull. Fac. Agric. Kagawa Univ.*, 18(2), 117, 1967.
7. Markley, K. S., *Soybeans and Soybean Products*, Vol. 1, Interscience, N.Y.,1950.
8. *Fats and Oils Situation*, U.S. Dept. Agric., FOS-255, November 1970.
9. *Soybean Dig.*, Blue Book Issue, 30(6), 1970.
10. *Official Grain Standards of the United States*, Grain Division, Consumer and Marketing Service, U.S. Dept. Agric., Hyattsville, Md., 1970.
11. Milner, M. and Geddes, W. F., *Cereal Chem.*, 23, 225, 1946.
12. Shotwell, O. L., Hesseltine, C. W., Burmeister, H. R., Kwolek, W. F., Shannon, G. M., and Hall, H. H., *Cereal Chem.*, 46, 454, 1969.
13. Hesseltine, C. W., Shotwell, O. L., Ellis, J. J., and Stubblefield, R. D., *Bacteriol. Rev.*, 30, 795, 1966.
14. Mitchell, H. H. and Beadles, J. R., *J. Nutr.*, 39, 463, 1949.
15. Norris, F., in *Bailey's Industrial Fats and Oils*, Swern, D., Ed., Interscience, N.Y., 1964, chap. 15.
16. Wilkens, W. F., Mattick, L.R., and Hand, D. B., *Food Technol.*, 21, 1630, 1967.
17. Oktomo, S., *Contemp. Manchuria*, 1(3), 83, 1937.
18. Duncan, I. J., *J. Am. Oil Chem. Soc.*, 25, 277, 1948; see Eldridge, A. C., *J. Am. Oil Chem. Soc.*, 46, 458A, 1969, for Bibliography on Solvents.

19. Moser, H. A., Evans, C. D., Campbell, R. E., Smith, A. K., and Cowan, J. C., *Cereal Sci. Today,* 12, 296, 1967.
20. McKinney, L. L., Weakley, F. B., Eldridge, A. C., Campbell, R. E., Cowan, J. C., Picken, J. C., Jr., and Biester, H. E., *J. Am. Chem. Soc.,* 79, 3932, 1957; 81, 909, 1959.
21. Ryder, J. W. and Sullivan, G. P., *J. Am. Oil Chem. Soc.,* 39, 263, 1962.
22. Lijinsky, W. and Robo, C. R., *Toxicol. Appl. Pharmacol.,* 3, 469, 1961.
23. Kruse, N. T., (Central Soya Co., Inc.), U.S. Patent 2,585,793, 1952.
24. Hayes, L. P. and Wolff, H., *J. Am. Oil Chem. Soc.,* 33, 440, 1956.
25. Norris, F., in *Bailey's Industrial Fats and Oils,* Swern, D., Ed., Interscience, N.Y., 1964, chap. 16.
26. Sullivan, F. E., *J. Am. Oil Chem. Soc.,* 45, 564A, 1968.
27. Cowan, J. C., *J. Am. Oil Chem. Soc.,* 43, 300A, 1966.
28. Singleton, W. A. and McMichael, C. E., *J. Am. Oil Chem. Soc.,* 32, 1, 1955.
29. Mattil, K. F., in *Bailey's Industrial Fats and Oils,* Swern, D., Ed., Interscience, N.Y., 1964, chap. 17.
30. Dutton, H. J., in *Progress in the Chemistry of Fats and Other Lipids, Part 3,* Vol. 9, Holman, R. T., Ed., Pergamon Press, London, 1968, chap. 9.
31. Dutton, H. J., Scholfield, C. R., Selke, E., and Rohwedder, W. K., *J. Catal.,* 10, 316, 1968.
32. Bailey, A. E., *J. Am. Oil Chem. Soc.,* 26, 644, 1949.
33. Beal, R. E. and Lancaster, E. B., *J. Am. Oil Chem. Soc.,* 31, 619, 1954.
34. Schmidt, H. J., *J. Am. Oil Chem. Soc.,* 47, 134, 1970.
35. Bailey, A. E., *J. Am. Oil Chem. Soc.,* 26, 166, 1949.
36. Bailey, A. E., (National Cylinder Gas Co.), U.S. Patent 2,691,665, 1954.
37. Zehnder, C. T. and McMichael, C. E., *J. Am. Oil Chem. Soc.,* 44, 478A, 1967.
38. Zehnder, C. T., paper presented at Am. Oil Chem. Soc. Short Course, September 1970.
39. Moser, H. A., Cooney, P. C., Evans, C. D., and Cowan, J. C., *J. Am Oil Chem. Soc.,* 43, 632, 1966.
40. Evans, C. D., Cooney, P. M., Scholfield, C. R., and Dutton, H. J., *J. Am. Oil Chem. Soc.,* 31, 295, 1954.
41. Moser, H. A., Evans, C. D., Cowan, J. C., and Kwolek, W. F., *J. Am. Oil Chem. Soc.,* 42, 30, 1965.
42. Stuckey, B. N., in *Handbook of Food Additives,* Furia, T. E., Ed., Chemical Rubber Co., Cleveland, 1968, chap. 5.
43. Dutton, H. J., Moser, H. A., and Cowan, J. C., *J. Am. Oil Chem. Soc.,* 24, 261, 1947.
44. McOsker, D. E., Mattson, F. H., Sweringen, H. B., and Kligman, A. M., *JAMA,* 180, 380, 1962.
45. Evans, C. D., Beal, R. E., McConnell, D. G., Black, L. T., and Cowan, J. C., *J. Am. Oil Chem. Soc.,* 41, 260, 1964.
46. Anon. *Consumer Rep.,* 29, 341, 1964.
47. National Soybean Processors Association, *Yearbook and Trading Rules,* Washington, 1969–1970, 34.
48. Agricultural Stabilization and Conservation Service, U.S. Dept. Agric., Soybean salad oil for export, Announcement PS-OM-28/64, 1964; Soybean salad oil, Announcement OC-9, 1970.
49. Jones, E. P., Scholfield, C. R., Davison, V. L., and Dutton, H. J., *J. Am. Oil Chem. Soc.,* 42, 727, 1965.
50. Scholfield, C. R., Davison, V. L. and Dutton, H. J., *J. Am. Oil Chem. Soc.,* 44, 648, 1967.
51. Mattil, K. F., in *Bailey's Industrial Fats and Oils,* Swern, D., Ed., Interscience, N.Y., 1964, chap. 8.
52. Joyner, N. T., *J. Am. Oil Chem. Soc.,* 30, 526, 1953.
53. Mattil, K. F., in *Bailey's Industrial Fats and Oils,* Swern, D., Ed., Interscience, N. Y., 1964, chap. 23.
54. Dutton, H. J., Schwab, A. W., Moser, H. A., and Cowan, J. C., *J. Am. Oil Chem. Soc.,* 25, 385, 1948.
55. Cowan, J. C., Cooney, P. M., and Evans, C. D., *J. Am. Oil Chem. Soc.,* 39, 6, 1962.
56. Goss, W. H., *Oil and Soap,* 23, 241, 1946.
57. Dutton H. J., Schwab, A. W., Moser, H. A., and Cowan, J. C., *J. Am. Oil Chem. Soc.,* 26, 441, 1949.
58. Evans, C. D., Schwab, A. W., Moser, H. A., Hawley, J. E., and Melvin, E. H., *J. Am. Oil Chem. Soc.,* 28, 68, 1951.
59. Durkee, M. M., *Ind. Eng. Chem.,* 28, 898, 1936.
60. Dutton, H. J., Lancaster, C. R., Evans, C. D., and Cowan, J. C., *J. Am. Oil Chem Soc.,* 28, 115, 1951.
61. Mattil, K. F., *J. Am. Oil Chem. Soc.,* 24, 243, 1947.
62. Hoffmann, R. L., Moser, H. A., Evans, C. D., and Cowan, J. C., *J. Am. Oil Chem. Soc.,* 39, 323, 1962.
63. Cowan, J. C., Evans, C. D., Moser, H. A., List, G. R., Koritala, S., Moulton, K. J., and Dutton, H. J., *J. Am. Oil Chem. Soc.,* 47, 470, 1970.
64. Smouse, T. H. and Chang, S. S., *J. Am. Oil Chem. Soc.,* 44, 509, 1967.
65. Hill, F. D. and Hammond, E. G., *J. Am. Oil Chem. Soc.,* 42, 1148, 1965.
66. Seals, R. G. and Hammond, E. G., *J. Am. Oil Chem. Soc.,* 43, 401, 1966; 47, 278, 1970.
67. Hoffman, G., *J. Am. Oil Chem. Soc.,* 38, 1, 1961.
68. Keppler, J. G., Schols, J. A., Feenstra, W. H., and Meijboom, P. W., *J. Am. Oil Chem. Soc.,* 42, 246, 1965.
69. Evans, C. D., List, G. R., Moser, H. A., and Cowan, J. C., *J. Am. Oil Chem. Soc.,* Abstr. 89, 45, 489A, 1968.
70. Iveson, H. T., *Soybean Dig.,* 21(8), 18, 1961.
71. Central Soya, Inc., Chicago, Ill., Lecithins, Technical Sales Manual.
72. Julian, P. L., Iveson, H. T., and McClelland, M. L. (Glidden Co.), U.S. Patent 2,629,662, 1953.
73. Smith, A. K. and Circle, S. J., *Ind. Eng. Chem.,* 30, 1414, 1938.
74. Burnett, R. S., in *Soybeans and Soybean Products,* Vol. 2, Markley, D., Ed., Interscience, N. Y., 1951, chap. 23.
75. Naismith, W. E. F., *Biochem. Biophys. Acta,* 16, 203, 1955.
76. Wolf, W. J. and Briggs, D. R., *Arch. Biochem. Biophys.,* 85, 186, 1959.

77. Wolf, W. J., *Baker's Dig.*, 43(5), 30, 1969.
78. Steiner, R. F. and Frattali, V., *J. Agric. Food Chem.*, 17, 513, 1969.
79. Wu, Y. V. and Scheraga, H. A., *Biochemistry*, 1, 698, 1962.
80. Fridman, C., Lis, H., Sharon, N., and Katchalski, E., *Arch. Biochem. Biophys.*, 126, 299, 1968.
81. Lis, H., Sharon, N., and Katchalski, E., *J. Biol. Chem.*, 241, 684, 1966.
82. Mitsuda, H., Yasumoto, K., Yamamoto, A., and Kusano, T., *Agric. Biol. Chem.*, 31, 115, 1967.
83. Gertler, A. and Birk, Y., *Biochem. J.*, 95, 621, 1965.
84. Koshiyama, I., *Cereal Chem.*, 45, 394, 1968.
85. Wolf, W. J., Babcock, G. E., and Smith, A. K., *Arch. Biochem. Biophys.*, 99, 265, 1962.
86. Lis, H., Fridman, C., Sharon, N., and Katchalski, E., *Arch. Biochem. Biophys.*, 117, 301, 1966.
87. Yamamoto, A., Yasumoto, K., and Mitsuda, H., *Agric. Biol. Chem.*, 34, 1169, 1970.
88. Christopher, J., Pistorius, E., and Axelrod, B., *Biochem. Biophys. Acta*, 198, 12, 1970.
89. Koshiyama, I., *Arch. Biochem. Biophys.*, 130, 370, 1969.
90. Wolf, W. J. and Sly, D. A., *Cereal Chem.*, 44, 653, 1967.
91. Catsimpoolas, N., Campbell, T. G., and Meyer, E. W., *Arch. Biochem. Biophys.*, 131, 577, 1969.
92. Hasegawa, K., Kusano, T., and Mitsuda, H., *Agric. Biol. Chem.*, 27, 878, 1963.
93. Obara, T. and Kimura, M., *J. Food Sci.*, 32, 531, 1967.
94. Wolf, W. J. and Sly, D. A., *Arch. Biochem. Biophys.*, 110, 47, 1965.
95. Puski, G. and Melnychyn, P., *Cereal Chem.*, 45, 192, 1968.
96. Shibasaki, K. and Okubo, K., *Tohoku J. Agric. Res.*, 16, 317, 1966.
97. Catsimpoolas, N., Leuthner, E., and Meyer, E. W., *Arch. Biochem. Biophys.*, 127, 338, 1968.
98. Briggs, D. R. and Wolf, W. J., *Arch. Biochem. Biophys.*, 72, 127, 1957.
99. Nash, A. M. and Wolf, W. J., *Cereal Chem.*, 44, 183, 1967.
100. Circle, S. J., Meyer, E. W., and Whitney, R. W., *Cereal Chem.*, 41, 157, 1964.
101. Wolf, W. J., Eldridge, A. C., and Babcock, G. E., *Cereal Chem.*, 40, 504, 1963.
102. Wolf, W. J. and Briggs, D. R., *Arch. Biochem. Biophys.*, 76, 377, 1958.
103. Catsimpoolas, N., Rogers, D. A., Circle, S. J., and Meyer, E. W., *Cereal Chem.*, 44, 631, 1967.
104. Catsimpoolas, N., *FEBS Lett.*, 4, 259, 1969.
105. Wolf, W. J., Rackis, J. J., Smith, A. K., Sasame, H. A., and Babcock, G. E., *J. Am. Chem. Soc.*, 80, 5730, 1958.
106. Catsimpoolas, N., Ekenstam, C., Rogers, D. A., and Meyer, E. W., *Biochem. Biophys. Acta*, 168, 122, 1968.
107. Wolf, W. J. and Tamura, T., *Cereal Chem.*, 46, 331, 1969.
108. Koshiyama, I., *Agric. Biol. Chem.*, 32, 879, 1968.
109. Koshiyama, I., *Agric. Biol. Chem.*, 34, 1815, 1970.
110. Kelley, J. J. and Pressey, R., *Cereal Chem.*, 43, 195, 1966.
111. Saio, K., Kamiya, M., and Watanabe, T., *Agric. Biol. Chem.*, 33, 1301, 1969.
112. Nash, A. M., Kwolek, W. F., and Wolf, W. J., *Cereal Chem.*, 48, 360, 1971.
113. McKinney, L. L., Sollars, W. F., and Setzkorn, E. A., *J. Biol. Chem.*, 178, 117, 1949.
114. Smith, A. K. and Rackis, J. J., *J. Am. Chem. Soc.*, 79, 633, 1957.
115. Belter, P. A. and Smith, A. K., *J. Am. Oil Chem. Soc.*, 29, 170, 1952.
116. Paulsen, T. M., Holt, K. E., and Anderson, R. E., *J. Am. Oil Chem. Soc.*, 37, 165, 1960.
117. American Oil Chemists' Society, *Official and Tentative Methods*, 3rd ed., Methods Ba 10-65 and Ba 11-65, 1970.
118. Smith, A. K., Belter, P. A., and Johnsen, V. L., *J. Am. Oil Chem. Soc.*, 29, 309, 1952.
119. Fukushima, D., *Bull. Agric. Chem. Soc. Jap.*, 23, 7, 1959.
120. Fukushima, D., *Bull. Agric. Chem. Soc. Jap.*, 23, 15, 1959.
121. Aoki, H. and Sakurai, M., *J. Agric. Chem. Soc. Jap.*, 42, 544, 1968.
122. Catsimpoolas, N. and Meyer, E. W., *Cereal Chem.*, 47, 559, 1970.
123. Watanabe, T. and Nakayama, O., *J. Agric. Chem. Soc. Jap.*, 36, 890, 1962.
124. Saio, K., Wakabayashi, A., and Watanabe, T., *J. Agric. Chem. Soc. Jap.*, 42, 90, 1968.
125. Rackis, J. J., Smith, A. K., Babcock, G. E., and Sasame, H. A., *J. Am. Chem. Soc.*, 79, 4655, 1957.
126. Rackis, J. J., Anderson, R. L., Sasame, H. A., Smith, A. K., and VanEtten, C. H., *J. Agric. Food Chem.*, 9, 409, 1961.
127. Central Soya Co., Inc., Chicago, Ill., Promosoy, soy protein concentrate, Technical Service Bulletin.
128. Central Soya Co., Inc., Chicago, Ill., Promine, an isolated soy protein, Technical Service Bulletin.
129. Saint, W. F., *Soybean Dig.*, 31(1), 32, 1970.
130. Watanabe, T., Industrial production of soybean foods in Japan, presented at Expert Group Meeting on Soya Bean Processing and Use, United Nations Industrial Development Organization, Peoria, Ill., November 1969.
131. Horan, F. E., Proc. Int. Conf. on Soybean Protein Foods, U. S. Dept. Agric. ARS-71-35, 129, 1967.
132. Kingsbaker, C. L., *J. Am. Oil Chem. Soc.*, 47, 458A, 1970.
133. Brekke, O. L., Mustakas, G. C., Raether, M. C., and Griffin, E. L., *J. Am. Oil Chem. Soc.*, 36, 256, 1959.
134. Mustakas, G. C., Albrecht, W. J., Bookwalter, G. N., McGhee, J. E., Kwolek, W. F., and Griffin, E. L., Jr., *Food Technol.*, 24, 1290, 1970.
135. O'Hara, J. B. and Schoepfer, A. E. (A. E. Staley Mfg. Co.), U.S. Patent 3,207,744, 1965.

136. Mustakas, G. C., Kirk, L. D., and Griffin, E. L., Jr., *J. Am. Oil Chem. Soc.,* 39, 222, 1962.
137. Moshy, R. J. (General Foods Corp.), U. S. Patent 3,126,286, 1964.
138. Sair, L. (The Griffith Laboratories Inc.), U. S. Patent 2,881,076, 1959.
139. McAnelly, J. K. (Swift and Co.), U. S. Patent 3,142,571, 1964.
140. Meyer, E. W., Proc. Int. Conf. on Soybean Protein Foods, U. S. Dept. Agric. ARS-71-35, 142, 1967.
141. Aspinall, G. O., Begbie, R., and McKay, J. E., *Cereal Sci. Today,* 12, 223, 1967.
142. Circle, S. J., Julian, P. L., and Whitney, R. W. (The Glidden Co.), U. S. Patent 2,881,159, 1959.
143. Wolf, W. J., Sly, D. A., and Kwolek, W. F., *Cereal Chem.,* 43, 80, 1966.
144. Eldridge, A. C., Wolf, W. J., Nash, A. M., and Smith, A. K., *J. Agric. Food Chem.,* 11, 323, 1963.
145. Nash, A. M., Eldridge, A. C., and Wolf, W. J., *J. Agric. Food Chem.,* 15, 102, 1967.
146. Eldridge, A. C. and Wolf, W. J., *Cereal Chem.,* 46, 344, 1969.
147. Eldridge, A. C., Hall, P. K., and Wolf, W. J., *Food Technol.,* 17, 1592, 1963.
148. Kuramoto, S., Westeen, R. W., and Keen, J. L. (General Mills, Inc.), U.S. Patent 3,177,079, 1965.
149. Eley, C. P., *Marketing and Transportation Situation,* U. S. Dept. Agric. ERS-388, 27, August 1968.
150. Wilding, M. D., *J. Am. Oil Chem. Soc.,* 47, 398, 1970.
151. Guy, E. J., Vettel, H. E., and Pallansch, M. J., *J. Dairy Sci.,* 52, 432, 1969.
152. Odell, A. D., *J. Inst. Can. Technol. Aliment.,* 2(4), A69, 1969.
153. Ziemba, J. V., *Food Eng.,* 43(1), 66, 1971.
154. Wolf, W. J., *J. Agric. Food Chem.,* 18, 969, 1970.
155. Johnson, D. W., *J. Am. Oil Chem. Soc.,* 47, 402, 1970.
156. Rakosky, J., Jr., *J. Agric. Food Chem.,* 18, 1005, 1970.
157. Inklaar, P. A. and Fortuin, J., *Food Technol.,* 23, 103, 1969.
158. Ralston Purina Co., Br. Patent 1,174,906, 1969.
159. Rock, H., Šipos, E. F., and Meyer, E. W., *Meat,* 32(6), 52, 1966.
160. Sair, L. and Melcer, I., Proc. Joint Symp. on Carbohydrate/Protein Interactions, Carbohydrate/Oilseeds Division, American Association of Cereal Chemists, June 22 to 23, 1970.
161. Wood, J. C., *Food Mfr.,* (Ingredient Survey), 11, 1967.
162. Tremple, L. G. and Meador, R. J., *Bakers' Dig.,* 32(4), 32, 1958.
163. Circle, S. J. and Johnson, D. W., in *Processed Plant Protein Foodstuffs,* Altschul, A. M., Ed., Academic Press, N.Y., 1958, chap. 15.
164. Pearson, A. M., Spooner, M. E., Hegarty, G. R., and Bratzler, L. J., *Food Technol.,* 19, 1841, 1965.
165. Anon., *Food Process. Mark. (Chic.),* 28(7), 48, 1967.
166. Ziemba, J. V., *Food Eng.,* 38(5), 82, 1966.
167. Turro, E. J. and Sipos, E., *Bakers' Dig.,* 42(6), 44, 1968.
168. Paulsen, T. M., *Food Technol.,* 15, 118, 1961.
169. Frank, S. S. and Circle, S. J., *Food Technol.,* 13, 307, 1959.
170. Rusoff, I. I., Ohan, W. J., and Long, C. L. (General Foods Corp.), U.S. Patent 3,047,395, 1962.
171. Ziemba, J. V., *Food Eng.,* 41(11), 72, 1969.
172. Thulin, W. W. and Kuramoto, S., *Food Technol.,* 21, 168, 1967.
173. Okumura, G. K. and Wilkinson, J. E., U.S. Patent 3,490,914, 1970.
174. Sakurai, Y., in *Food Technology the World Over,* Vol. 2, Peterson, M. S. and Tressler, D. K., Eds., Avi Publishing, Westport, Conn., 1965, chap 23.
175. Aoki, H., *J. Agric. Chem. Soc. Jap.,* 39, 270, 1965.
176. Mizrahi, S., Zimmermann, G., Berk, Z., and Cogan, U., *Cereal Chem.,* 44, 193, 1967.
177. Turro, E. J. and Sipos, E., *Baker's Dig.,* 44(1), 58, 1970.
178. Anon., *Meat,* 36(8), 24, 1970.
179. Westeen, R. W. and Kuramoto, S. (General Mills, Inc.), U.S. Patent, 3,118,959, 1964.
180. Unilever, N. V., Neth. Patent Appl. 69, 12,222, 1970; C. A. 73,34025j, 1970.
181. Ishler, N. H., MacAllister, R. V., Szczesniak, A. S., and Engel, E.(General Foods Corp.), U.S. Patent 3,093,483, 1963.
182. Corey, H., *CRC Crit. Rev. Food Technol.,* 1, 161, 1970.
183. Coleman, R. J. and Creswick, N. S. (T. J. Lipton, Inc.), U.S. Patent 3,253,931, 1966.
184. Kies, M. W., Haining, J. L., Pistorius, E., Schroeder, D. H., and Axelrod, B., *Biochem. Biophys. Res. Commun.,* 36, 312, 1969.
185. Osborne, T. B. and Mendel, L. B., *J. Biol. Chem.,* 32, 369, 1917.
186. Mickelsen, O. and Yang, M. G., *Fed. Proc.,* 25, 104, 1966.
187. Liener, I. E. and Kakade, M. L., in *Toxic Constituents of Plant Foodstuffs,* Liener, I. E., Ed., Academic Press, N.Y., 1969, chap. 2.
188. Kunitz, M., *J. Gen. Physiol.,* 30, 291, 1947.
189. Birk, Y., *Biochem. Biophys. Acta,* 54, 378, 1961.
190. Gertler, A., Birk, Y., and Bondi, A., *J. Nutr.,* 91, 358, 1967.
191. Frattali, V., *J. Biol. Chem.,* 244, 274, 1969.

192. Birk, Y., Gertler, A., and Khalef, S., *Biochem. J.*, 87, 281, 1963.
193. Yamamoto, M. and Ikenaka, T., *J. Biochem.* (Tokyo), 62, 141, 1967.
194. Gorrill, H. D. L. and Thomas, J. W., *J. Nutr.*, 92, 215, 1967.
195. Hooks, R. D., Hays, V. W., Speer, V. C., and McCall, J. T., *J. Anim. Sci.*, 24, 894, 1965.
196. Lyman, R. L. and Lepkovsky, S., *J. Nutr.*, 62, 269, 1957.
197. Khayambashi, H. and Lyman, R. L., *Am. J. Physiol.*, 217, 646, 1969.
198. Melmed, R. N. and Bouchier, I. A. D., *Gut*, 10, 973, 1969.
199. Travis, J. and Roberts, R. C., *Biochemistry*, 8, 2884, 1969.
200. Roberts, R. C., personal communication, 1970.
201. Rackis, J. J., *Fed. Proc.*, 24, 1488, 1965.
202. Rackis, J. J., *Food Technol.*, 20, 1482, 1966.
203. Albrecht, W. J., Mustakas, G. C., and McGhee, J. E., *Cereal Chem.*, 43, 400, 1966.
204. Kotter, L., Palitzsch, A., Belitz, H.-D., and Fischer, K.-H., *Die Fleischwirtschaft*, 50, 1063, 1970.
205. Eldridge, A. C., Anderson, R. L., and Wolf, W. J., *Arch. Biochem. Biophys.*, 115, 495, 1966.
206. Schingoethe, D. J., Aust, S. D., and Thomas, J. W., *J. Nutr.*, 100, 739, 1970.
207. Jaffe, W. G., in *Toxic Constituents of Plant Foodstuffs*, Liener, I. E., Ed., Academic Press, N.Y., 1969, chap 3.
208. Liener, I. E. and Rose, J. E., *Proc. Soc. Exp. Biol. Med.*, 83, 539, 1953.
209. Liener, I. E., *Arch. Biochem. Biophys.*, 54, 223, 1955.
210. Liener, I. E., *J. Biol. Chem.*, 233, 401, 1958.
211. Liener, I. E., and Pallansch, M. J., *J. Biol. Chem.*, 197, 29, 1952.
212. Liener, I. E., *J. Nutr.*, 49, 527, 1953.
213. Birk, Y. and Gertler, A., *J. Nutr.*, 75, 379, 1961.
214. Jaffe, W. G., *Arzneim-Forsch.*, 10, 1012, 1960.
215. Gestetner, B., Birk, Y., Bondi, A., and Tencer, Y., *Phytochemistry*, 5, 803, 1966.
216. Ishaaya, I., Birk, Y., Bondi, A., and Tencer, Y., *J. Sci. Food Agric.* 20, 433, 1969.
217. Ishaaya, I. and Birk, Y., *J. Food Sci.*, 30, 118, 1965.
218. Wolf, W. J. and Thomas, B. W., *J. Am. Oil Chem. Soc.*, 47, 86, 1970.
219. Wolf, W. J. and Thomas, B. W., *J. Chromatogr.*, 56, 281, 1971.
220. Birk, Y., in *Toxic Constituents of Plant Foodstuffs*, Liener, I. E., Ed., Academic Press, N. Y., 1969, chap. 7.
221. Food and Agriculture Organization of the United Nations, Protein Requirements, FAO Nutrition Meetings Report No. 37, 1965.
222. Theuer, R. C. and Sarett, H. P., *J. Agric. Food Chem.*, 18, 913, 1970.
223. Wilding, M. D., Alden, D. E., and Rice, E. E., *Cereal Chem.*, 45, 254, 1968.
224. Rice, E. E., *J. Am. Oil Chem. Soc.*, 47, 408, 1970.
225. Cahill, W. M., Schroeder, L. J., and Smith, A. H., *J. Nutr.*, 28, 209, 1944.
226. Lewis, J. H. and Taylor, F. H. L., *Proc. Soc. Exp. Biol. Med.*, 64, 85, 1947.
227. Huang, P.-C., Tung, T.-C., Lue, H.-C., and Wei, H.-Y., *J. Formosan Med. Assoc.*, 64, 591, 1965.
228. Schneider, D. L. and Sarett, H. P., *J. Nutr.*, 98, 279, 1969.
229. Althoff, J. D., *Med. Klin.*, 65, 1204, 1970.
230. Iriarte, B. J. R. and Barnes, R. H., *Food Technol.*, 20, 835, 1966.
231. Hamdy, M. M., Case, W. H., Cole, M. S., and Horan, F. E. Protein Advisory Group FAO/WHO/UNICEF, Geneva, Document 2.24/1, September 1969.
232. Longenecker, J. B., Martin, W. H., and Sarett, H. P., *J. Agric. Food Chem.*, 12, 411, 1964.
233. Rackis, J. J., Smith, A. K., Nash, A. M., Robbins, D. J., and Booth, A. N., *Cereal Chem.*, 40, 531, 1963.
234. Bressani, R., Viteri, F., Elias, L. G., de Zaghi, S., Alvarado, J., and Odell, A. D., *J. Nutr.*, 93, 349, 1967.
235. Cogan, U., Yaron, A., Berk, Z., and Zimmermann, G., *J. Agric. Food Chem.*, 16, 196, 1968.
236. Koury, S. D. and Hodges, R. E., *J. Am. Diet. Assoc.*, 52, 480, 1968.
237. Standal, B. R., *J. Nutr.*, 81, 279, 1963.
238. Hackler, L. R., Hand, D. B., Steinkraus, K. H., and Van Buren, J. P., *J. Nutr.*, 80, 205, 1963.
239. Smith, A. K., Rackis, J. J., Hesseltine, C. W., Smith, M., Robbins, D. J., and Booth, A. N., *Cereal Chem.*, 41, 173, 1964.
240. Gray, W. D., *CRC Crit. Rev. Food Technol.*, 1, 225, 1970.
241. Watanabe, T., Japanese Patent 70 16,773, June 10, 1970; C. A. 74, 12030a, 1971.
242. Anon., *Chem. Eng. News*, 48(43), 11, 1970.
243. Turro, E. J. (Central Soya Co., Inc.), U.S. Patent 3,529,969, 1970.
244. Hartman, W. E. (Worthington Foods, Inc.), U.S. Patent 3,320,070, 1967.
245. Corkern, R. S. and Dwoskin, P. B., Consumer acceptance of a new bacon substitute, U.S.D.A., ERS 454, October 1970.
246. Smith, R. E., Proc. Joint Symp. Carbohydrate/Protein Interactions, Carbohydrate/Oilseeds Division, American Association of Cereal Chemists, June 22 to 23, 1970.
247. Anon., *Chem. Eng. News*, 48(33), 36, 1970.
248. McCormick, R. D. and Beck, K. M., *Food Prod. Dev.*, 2(1), 26, 1968.

249. Fujimaki, M., Yamashita, M., Okazawa, Y., and Arai, S., *J. Food Sci.*, 35, 215, 1970.
250. Sair, L. (*Griffith Laboratories, Inc.,*), U.S. Patent 3,391,001, 1968.
251. Manley, C. H. and Fagerson, I. S., *J. Food Sci.*, 35, 286, 1970.
252. Manley, C. H. and Fagerson, I. S., *J. Agric. Food Chem.*, 18, 340, 1970.
253. Kalbrener, J. E., Eldridge, A. C., Moser, H. A., and Wolf, W. J., *Cereal Chem.*, 48, 595, 1971.
254. Iden, R., Livingston, R., and Watson, C. A., *Cereal Chem.*, 47, 43, 1970.
255. Arai, S., Koyanagi, O., and Fujimaki, M., *Agric. Biol. Chem.*, 31, 868, 1967.
256. Fujimaki, M., Arai, D., Kirigaya, N., and Sakurai, Y., *Agric. Biol. Chem.*, 29, 855, 1965.
257. Arai, S., Suzuki, H., Fujimaki, M., and Sakurai, Y., *Agric. Biol. Chem.*, 30, 364, 1966.
258. Arai, S., Suzuki, H., Fujimaki, M., and Sakurai, Y., *Agric. Biol. Chem.*, 30, 863, 1966.
259. Sessa, D. J., Honig, D. H., and Rackis, J. J., *Cereal Chem.*, 46, 675, 1969.
260. Arai, S., Noguchi, M., Yamashita, M., Kato, H., and Fujimaki, M., *Agric. Biol. Chem.*, 34, 1569, 1970.
261. Van der Meer, J. H. H. and Spaans, J., Abstr. 55th Ann. Meet., American Association of Cereal Chemists, October 18 to 22, 1970.
262. Mattick, L. R. and Hand, D. B., *J. Agric. Food Chem.*, 17, 15, 1969.
263. Badenhop, A. F. and Wilkins, W. F., *J. Am. Oil Chem. Soc.*, 46, 179, 1969.
264. Wilkens, W. F. and Lin, F. M., *J. Agric. Food Chem.*, 18, 333, 1970.
265. Kon, S., Wagner, J. R., Guadagni, D. G., and Horvat, R. J., *J. Food Sci.*, 35, 343, 1970.
266. Honig, D. H., Sessa, D. J., Hoffmann, R. L., and Rackis, J. J., *Food Technol.*, 23, 803, 1969.
267. Arai, S., Noguchi, M., Yamashita, M., Kato, H., and Fujimaki, M., *Agric. Biol. Chem.*, 34, 1338, 1970.
268. Eldridge, A. C., Kalbrener, J. E., Moser, H. A., Honig, D. H., Rackis, J. J., and Wolf, W. J., *Cereal Chem.*, 48, 640, 1971.
269. Steggerda, F. R., Richards, E. A., and Rackis, J. J., *Proc. Soc. Exp. Biol. Med.*, 121, 1235, 1966.
270. Rackis, J. J., Honig, D. H., Sessa, D. J., and Steggerda, F. R., *J. Agric. Food Chem.*, 18, 977, 1970.
271. Cristofaro, E., Mottu, F., and Wuhrmann, J. J., Study of the effect of stachyose and raffinose on the flatulence activity of soy milk, presented at 3rd Int. Cong. Food Sci. Technol., Washington, August 9 to 14, 1970.
272. Arai, S., Noguchi, M., Kurosawa, S., Kato, H., and Fujimaki, M., *J. Food Sci.*, 35, 392, 1970.
273. Melnychyn, P. and Stapley, R. B. (Carnation Co.), S. African Patent 687,706, 1969.
274. Anon., *Fed. Reg.*, 34(194), 15657, October 8, 1969.
275. Martin, R. E., *Soybean Dig.*, 31(1), 15, 1970.
276. Rackis, J. J., Honig, D. H., Sessa, D. J., and Cavins, J. F., *J. Food Sci.*, 36, 10, 1971.
277. Morse, W. J., in *Soybeans and Soybean Products,* Vol. 1, Markley, K. S., Ed., Interscience, N.Y., 1950, chap. 1.
278. Hymowitz, T., *Econ. Bot.*, 24, 408, 1972.
279. Kromer, G. W., *Fats and Oils Situation*, U.S. Dept. Agric., FOS-267, April 1973.
280. Kromer, G. W., Palm oil in the world's fat and oil economy, Presented at Fall Meet., American Oil Chemists' Society, September 30, 1974.
281. Sullivan, F., *Food Eng.*, 46(5), 79, 1974.
282. Thomason, F. G., *Fats and Oils Situation*, U.S. Dept. Agric., FOS-275, November 1974.
283. Hesseltine, C. W., de Camargo, R., and Rackis, J. J., *Nature*, 200, 1226, 1963.
284. *Fats and Oils Situation*, U.S. Dept. Agric., FOS-275, November 1974.
285. *Fats and Oils Situation*, U.S. Dept. Agric., FOS-274, October 1974.
286. Kromer, G. W., *Fats and Oils Situation*, U.S. Dept. Agric., FOS-272, April 1974.
287. Anon., *Soybean Dig.*, 34(13), 8, 1974.
288. Anon., *Soybean Dig.*, 34(8), 14, 1974.
289. Anon., *Soybean Dig.*, 34(4), 6, 1974.
290. Anon., *Soybean Dig.*, 34(11), 12, 1974.
291. Brim, C. A., and Young, M. F., *Crop Sci.*, 11, 564, 1974.
292. Hymowitz, T., Collins, F. I., Panczmes, J., and Walker, W. M., *Agron. J.*, 64, 613, 1972.
293. Hymowitz, T., Walker, W. M., Collins, F. I., and Panczmes, J., *Soil Sci. Plant Anal.*, 3, 367, 1972.
294. Mies, D. W. and Hymowitz, T., *Bot. Gaz.*, 134, 121, 1973.
295. Bean, G. A., Schillinger, J. A., and Klarman, W. L., *Appl. Microbiol.*, 24, 437, 1972.
296. Lineberry, D. P. and Dudrow, F., U.S. Patent 3,693,322, 1972.
297. Sullivan, F. E., *Chem. Eng.*, 81(8), 56, 1974.
298. *Tenox® TBHQ Antioxidant for Oils, Fats and Fat Containing Foods,* Eastman Chemical Products, Inc., Publication ZG-201A, July 1973.
299. Code of Federal Regulations 21, 121.1244, 1973 (see *Fed. Reg.*, 37, 25357, November 30, 1972).
300. Official and Tentative Methods of the American Oil Chemists Society, 3rd ed., Cd 12-57, 1973.
301. Sherwin, E. R. and Thompson, J. W., *Food Technol.*, 21, 912, 1967.
302. List, G. R., Evans, C. D., Beal, R. E., Black, L. T., Moulton, K. J., and Cowan, J. C., *J. Am. Oil Chem. Soc.*, 51, 239, 1974.
303. Moulton, K. J., Beal, R. E., and Griffin, E. L., *J. Am. Oil Chem. Soc.*, 50, 450, 1973.

304. Cowan, J. C., Koritala, S., Warner, K., List, G. R., Moulton, K. J., and Evans, C. D., *J. Am. Oil Chem. Soc.*, 50, 132, 1973.
305. Vigneron, P. Y., Koritala, S., Butterfield, R. O., and Dutton, H. J., *J. Am. Oil Chem. Soc.*, 49, 371, 1972.
306. Moulton, K. J., Moore, D. J., and Beal, R. E., *J. Am. Oil Chem. Soc.*, 46, 662, 1969.
307. Beal, R. E., Moulton, K. J., Moser, H. A., and Black, L. T., *J. Am. Oil Chem. Soc.*, 46, 498, 1969.
308. Krober, O. A. and Collins, F. I., *J. Am. Oil Chem. Soc.*, 25, 296, 1948.
309. Sanders, J. H., *J. Am. Oil Chem. Soc.*, 21, 357, 1944.
310. Anon., *Soybean Dig.*, 32 (2), 7, 1971.
311. Evans, C. D., List, G. R., Beal, R. E., and Black, L. T., *J. Am. Oil Chem. Soc.*, 51, 444, 1974.
312. List, G. R., Evans, C. D., Warner, K., Boundy, B. K., and Beal, R. E., *J. Am. Oil Chem. Soc.*, 51, 531A, 1974; Abstr 186, Fall Meet., American Oil Chemists' Society, September 30, 1974.
313. Meijboom, P. W. and Stroink, J. B. A., *J. Am. Oil Chem. Soc.*, 49, 555, 1972.
314. Badings, H. T., *J. Am. Oil Chem. Soc.*, 50, 334, 1973.
315. Haverkamp Begemann, P. and Koster, J. C., *Nature*, 202, 552, 1964.
316. Lockmiller, N. R., in *Fabricated Foods*, Inglett, G. E., Ed., Avi Publishing, Westport, Conn., 1975, chap. 5.
317. Lockmiller, N. R., *Cereal Sci. Today*, 18, 77, 1973.
318. Nelson, A. I., Wei, L. S., and Steinberg, M. P., *Soybean Dig.*, 31(3) 32, 1971.
319. Shemer, M., Wei, L. S., and Perkins, E. G., *J. Food Sci.*, 38, 112, 1973.
320. Mustakas, G. C., Albrecht, W. J., Bookwalter, G. N., Sohns, V. E., and Griffin, E. J., Jr., *Food Technol.*, 25, 534, 1971.
321. Mustakas, G. C., *Cereal Sci. Today*, 19, 62, 1974.
322. Nielsen, K., *J. Am. Oil Chem. Soc.*, 37, 217, 1960.
323. Rackis, J. J., Eldridge, A. C., Kalbrener, J. E., Honig, D. H., and Sessa, D. J., *AICHE Symp. Ser.*, 69(132), 5, 1973.
324. Steinkraus, K. H. (Cornell Research Foundation, Inc.), U.S. Patent 3,721,569, 1973.
325. DeLapp, D. F. (American Cyanamid Co.), U.S. Patent 3,762,929, 1973.
326. Smith, A. K., Nash, A. M. Eldridge, A. C., and Wolf, W. J., *J. Agric. Food Chem.*, 10, 302, 1962.
327. Schweiger, R. G. and Muller, S. A., (Grain Processing Corp.), U.S. Patent 3,714,210, 1973.
328. Hayes, L. P. and Simms, R. P. (A. E. Staley Manufacturing Co.), U.S. Patent 3,734,901, 1973.
329. Glombert, J. D. (Unilever N.V.), Ger. Offen. 2,355, 892, 1974; C.A. 81, 48762, 1974.
330. Hoer, R. A. and Calvert, F. E. (Ralston Purina Co.), U.S. Patent 3,649,293, 1972.
331. Miller, D. M. and Wilding, M. D. (Swift and Co.), U.S. Patent 3,723, 407, 1973.
332. Smith, A. K., Rackis, J. J., Isnardi, P., Cartter, J. L., and Krober, O. A., *Cereal Chem.* 43, 261, 1966.
333. Wolf, W. J. and Baker, F. L., *Cereal Sci. Today*, 17, 124, 1972.
334. Hoer, R. A., Frederiksen, C. W., and Hawley, R. L., (Ralston Purina Co.), U.S. Patent 3,694,221, 1972.
335. Gutcho, M., *Textured Foods and Allied Products*, Food Technol. Rev. No. 1, Noyes Data Corp., Park Ridge, N.J., 1973.
336. Boyer, R. A., Schulz, A. A., Oborsh, E. V., and Brown, A. V., (Ralston Purina Co.), U.S. Patent 3,662,673, 1972.
337. Strommer, P. K., (General Mills, Inc.), U.S. Patent 3,730,729, 1973.
338. Strommer, P. K. and Beck, C. I. (General Mills, Inc.), U.S. Patent 3,754,926, 1973.
339. Tombs, M. P. (Unilever Ltd.), Br. Patent 1,265,661, 1972:
340. DeGroot, A. P. and Slump, P., *J. Nutr.*, 98, 45, 1969.
341. Frederiksen, C. W. and Heusdens, W. (Ralston Purina Co.), U.S. Patent 3,662,671, 1972.
342. Hoer, R. A. (Ralston Purina Co.), U.S. Patent 3,662,672, 1972.
343. Hashizume, K., Kosaka, K., Koyama, E., and Watanabe, T., *J. Food Sci. Technol.* (Tokyo), 21, 136, 1974.
344. Hashizume, K., Nakamura, N., and Watanabe, T., *J. Food Sci. Technol.* (Tokyo), 21, 141, 1974.
345. Briskey, E. J., in *Evaluation of Novel Protein Products*, Vol. 14, Proc. Int. Biol. Programme and Wenner-Gren Center Symp., Bender, A. E., Kihlberg, R., Löfquist, B., and Munck, L., Eds., Pergamon Press, N. Y., 1970.
346. Hermansson, A. -M., in *Proteins in Human Nutrition*, Porter, J. W. G. and Rolls, B. A., Eds., Academic Press, N. Y. 1973, chap. 27.
347. van Megen, W. H., *J. Agric. Food Chem.*, 22, 126, 1974.
348. Hagenmaier, R., *J. Food Sci.*, 37, 965, 1972.
349. Hermansson, A. -M., *Lebensm. -Wiss. Technol.*, 5, 24, 1972.
350. Baumann, H., *Fette Seifen Anstrichm.*, 68, 741, 1966.
351. Fleming, S. E., Sosulski, F. W., Kilara, A., and Humbert, E. S., *J. Food Sci.*, 39, 188, 1974.
352. Smith, G. C., Juhn, H., Carpenter, Z. L., Mattil, K. F., and Cater, C. M., *J. Food Sci.*, 38, 849, 1973.
353. Crenwelge, D. D., Dill, C. W., Tybor, P. T., and Landmann, W. A., *J. Food Sci.*, 39, 175, 1974.
354. Wu, L. C. and Bates, R. P., *J. Food Sci.*, 37, 36, 1972.
355. Wu, L. C. and Bates, R. P., *J. Food Sci.*, 37, 40, 1972.
356. Wu, L. C. and Bates, R. P., *J. Food Sci.*, 38, 783, 1973.
357. Cumming, D. B., Stanley, D. W., and DeMan, J. M., *Can. Inst. Food Sci. Technol. J.*, 5, 124, 1972.
358. Cumming, D. B., Stanley, D. W., and DeMan, J. M., *J. Food Sci.*, 38, 320, 1973.
359. Stanley, D. W., Cumming, D. B., and DeMan, J. M., *Can. Inst. Food Sci. Technol. J.*, 5, 118, 1972.

360. Chiang, J. P. C. and Sternberg, M., *Cereal Chem.*, 51, 465, 1974.
361. Saio, K. and Watanabe, T., *J. Food Sci.*, 38, 1139, 1973.
362. Saio, K., Sato, I., and Watanabe, T., *J. Food Sci.*, 39, 777, 1974.
363. Liener, I. E., in *Soybeans: Chemistry and Technology. Proteins*, Vol 1, Smith, A. K. and Circle S. J., Eds., Avi Publishing, Westport, Conn., 1972, chap. 7.
364. Rackis, J. J., in *Soybeans: Chemistry and Technology. Proteins*, Vol. 1, Smith, A. K. and Circle, S. J., Eds., Avi Publishing, Westport, Conn., 1972, chap. 6.
365. Rackis, J. J., *J. Am. Oil Chem. Soc.*, 51, 161A, 1974.
366. Kakade, M. L., Simons, N. R., Liener, I. E., and Lambert, J. W., *J. Agric. Food Chem.*, 20, 87, 1972.
367. Hymowitz, T. and Hadley, H. H., *Crop Sci.*, 12, 197, 1972.
368. Hymowitz, T., *Crop Sci.*, 13, 420, 1973.
369. Clark, R. W., Mies, D. W., and Hymowitz, T., *Crop Sci.*, 10, 486, 1970.
370. Yen, J. T., Jensen, A. H., Hymowitz, T., and Baker, D. H., *Poult. Sci.*, 52, 1875, 1973.
371. Kakade, M. L., Hoffa, D. E., and Liener, I. E., *J. Nutr.*, 103, 1772, 1973.
372. Kakade, M. L., Rackis, J. J., McGhee, J. E., and Puski, G., *Cereal Chem.*, 51, 376, 1974.
373. Rackis, J. J., McGhee, J. E., and Booth, A. N., *Cereal Chem.*, 52, 85, 1975.
374. Patten, J. R., Richards, E. A., and Pope, H., *Proc. Soc. Exp. Biol. Med.*, 137, 58, 1971.
375. Patten, J. R., Richards, E. A., and Wheeler, J., *Life Sci.*, 10(2), 145, 1971.
376. Nordal, J. and Fossum, K., *Z. Lebensm.-Unters.-Forsch.*, 154, 144, 1974.
377. Graham, G. G., Morales, E., Acevedo, G., Placko, R. P., and Cordano, A., *Am. J. Clin. Nutr.*, 24, 416, 1971.
378. Graham, G. G., Baertl, J. M., Placko, R. P., and Morales, E., *Am. J. Clin. Nutr.*, 26, 491, 1973.
379. Graham, G. G., Baertl, J. M., Placko, R. P., and Cordano, A., *Am. J. Clin. Nutr.*, 25, 875, 1972.
380. Beghin, I., DeMello, A. V., Costa, T., Monteiro, E., Lucena, M. A., and Varela, R., *Am. J. Clin. Nutr.*, 26, 246, 1973.
381. Del Valle, F. R., and Perez-Villasenor, J., *J. Food Sci.*, 39, 244, 1974.
382. Bressani, R., Murillo, B., and Elias, L. G., *J. Food Sci.*, 39, 577, 1974.
383. Kies, C. and Fox, H. M., *J. Food Sci.*, 36, 841, 1971.
384. Kies, C. and Fox, H. M., *J. Food Sci.*, 38, 1211, 1973.
385. Poullain, B., Guisard, D., and Debry, G., *Nutr. Metab.*, 14, 298, 1972.
386. Morse, E. H., Merrow, S. B., Keyser, D. E., and Clark, R. P., *Am. J. Clin. Nutr.*, 25, 912, 1972.
387. Graham, G. G., Placko, R. P., Morales, E., Acevedo, G., and Cordano, A., *Am. J. Dis. Child.*, 120, 419, 1970.
388. Fomon, S. J., Thomas, L. N., Filer, L. J., Jr., Anderson, T. A., and Bergmann, K. E., *Acta Paediatr. Scand.*, 62, 33, 1973.
389. Woodard, J. C. and Short, D. D., *J. Nutr.*, 103, 569, 1973.
390. Cowan, J. C., Rackis, J. J., and Wolf, W. J., *J. Am. Oil Chem. Soc.*, 50, 426A, 1973.
391. Maga, J. A., *J. Agric. Food Chem.*, 21, 864, 1973.
392. Wolf, W. J., *J. Agric. Food Chem.*, 23, 136, 1975.
393. Grosch, W. and Schwencke, D., *Lebensm.-Wiss. Technol.*, 2, 109, 1969.
394. Kalbrener, J. E., Warner, K., and Eldridge, A. C., *Cereal Chem.*, 51, 406, 1974.
395. Sessa, D. J., Warner, K., and Honig, D. H., *J. Food Sci.*, 39, 69, 1974.
396. Cotton, R. H., *J. Am. Oil Chem. Soc.*, 51, 116A, 1974.
397. Singleton, A. D. and Robertson, R. G., *Baker's Dig.*, 48(1), 46, 1974.
398. Anon., *Food Process.*, 34(8), 32, 1973.
399. Anon., *Food Process.*, 35(5), 18, 1974.
400. Tsen, C. C. and Hoover, W. J., *Cereal Chem.*, 50, 7, 1973.
401. Tsen, C. C., Peters, E. M., Schaffer, T., and Hoover, W. J., *Baker's Dig.*, 47(4), 34, 1973.
402. Bird, K., *Cereal Sci. Today*, 19, 226, 1974.
403. Wolford, K. M., *J. Am. Oil Chem. Soc.*, 51, 131A, 1974.
404. Iammartino, N. R., *Chem. Eng.*, 81(16), 50, 1974.
405. Rosenfield, D. and Hartman, W. E., *J. Am. Oil Chem. Soc.*, 51, 91A, 1974.
406. Horan, F. E., in *New Protein Foods*, Vol. 1A, Altschul, A. M., Ed., Academic Press, N.Y., 1974, chap. 8.
407. Claus, W. S., *J. Am. Oil Chem. Soc.*, 51, 197A, 1974.
408. Wodicka, V. O., *J. Am. Oil Chem. Soc.*, 51, 101A, 1974.
409. Mussman, H. C., *J. Am. Oil Chem. Soc.*, 51, 104A, 1974.
410. Czarnecki, J. N., *J. Am. Oil Chem. Soc.*, 51, 110A, 1974.
411. Ward, A. G., *J. Am. Oil Chem. Soc.*, 51, 107A, 1974.
412. Watanabe, T., *J. Am. Oil Chem. Soc.*, 51, 111A, 1974.
413. Hutt, P. B., in *New Protein Foods*, Vol. 1A, Altschul, A. M., Ed., Academic Press, N.Y., 1974, chap. 10.
414. Smith, P. R., *10th Anniv., Mini-symp., Meat and Meat Products*, Inst. Food Sci. Technol., U. K., March 7 to 8, 1974, p 21-29.
415. Formo, M. W., Honold, G. R., and MacLean, D. B., *J. Assoc. Off. Agric. Chem.*, 57, 841, 1974.
416. Lee, Y. B., Rickansrud, D. A., Hagberg, E. C., Briskey, E. J., and Greaser, M. L., *J. Food Sci*, 40, 380, 1975.

INDEX

A

Acreage, soybean, 5-6, 72-73
Aflatoxin in soybeans, 9, 76
Alkali treatment, effect on soybean protein, 29, 86, 87
Amino acid composition of soybean proteins, 33, 34, 55, 56
Antifoaming agent in soybean oil, 17, 23
Antinutritional factors in soybeans, 51-54, 75, 76
Antioxidants, 17, 77
    tertiary-butylhydroquinone (TBHQ), 77

B

Baked goods, soybean proteins in, 42, 43, 44, 45, 46, 51, 60, 61, 62, 88, 89
Beverage products, soybean protein in, 61, 64, 65, 80
Bitterness of oxidized phosphatidyl choline, 88
Breakfast cereals, soybean proteins in, 42, 61, 64

C

Carbohydrates, 3, 75
Catalysts for hydrogenation, 14, 15, 77, 78
Citric acid (metal inactivating agent), 17, 18, 22
Composition of soybeans, 3, 5
Concentrates, protein, see Soy protein concentrates
Corn-soy-milk blended food (CSM), 42, 85, 86
Crop production, see Seed, production

D

Defatting of soybeans, see Extraction of oil
Degumming of soybean oil, 12
Denaturation of proteins, 30-33
Desolventizing of flakes, 12, 35, 36, 37, 38
Dietary foods, soybean proteins in, 61, 65

E

Emulsifying properties of soybean proteins, 43, 44, 84
Essential amino acids, 33, 34, 55, 56
Essential fatty acids, see Polyunsaturates and P/S ratio
Extraction of oil, 11-12
Extraction of residual lipids from defatted flakes, 67, 81
Extruded soy flour products, see Soy flours and grits
Extrusion cooking for full-fat soy flour, 37, 39, 80

F

Fibers, see Spun protein fibers
Field damaged soybeans, 14, 78, 79
Flash desolventizer, 35, 37
Flatus factors, 67, 75
Flavor of protein products, 65-67, 87, 88
Flavor stability of soybean oil, 22-24, 77-79
Flours and grits, see Soy flours and grits
Food uses of soybean proteins, 59-65, 80, 88, 89, 90
  baked goods, 42, 43, 44, 45, 46, 51, 60, 61, 62, 88
  beverages, 61, 64, 65, 80
  breakfast cereals, 42, 61, 64
  dietary foods, 61, 65
  infant foods, 42, 57, 61, 64, 85, 86
  instant breakfast items, 89
  meat products and analogs, 42, 43, 44, 46, 47, 62, 63, 64, 88, 89
  miscellaneous, 65
  oriental foods, 58, 59, 60
  snack foods, 61, 65, 89, 90
Full-fat soy flour, 37, 39, 60, 80
Functional properties of soybean proteins, 43-51, 67, 68, 83-85
  adhesion, 44, 50
  aeration, 44, 51
  cohesion, 44, 50
  color control, 44, 51
  dough formation, 44, 48, 50
  elasticity, 44, 50
  emulsification, 43, 44, 84
  fat absorption, 43, 44, 45
  film formation, 44, 50, 51, 84
  gelation, 31, 44, 46
  solubility, 25, 26, 29, 30, 31, 40, 41, 67, 68, 83
  texture, 44, 46, 47, 48, 84
  viscosity, 44, 84
  water absorption, 44, 45, 46, 83

G

Gel filtration of soybean proteins, 27, 28
Gelation, 31, 44, 46
Grade standards, 7

H

Hemagglutinins, 53

I

Infant foods, soybean proteins in, 42, 57, 61, 64, 85, 86
Information, sources, 71
  proceedings of meetings, 71
  recent books, 71
Instant breakfast items, soybean proteins in, 89
Isolates, protein, see Soy protein isolates

K

Kinako, see Oriental soybean foods

L

Lecithin, 12, 24, 25
Lecithinated soy flours, 37
Legal aspects of soybean proteins in foods, 62, 63, 68, 90
Linoleic acid, 20, 78, Author's introduction
Linolenic acid, 20, 22, 23, 78
Lipids, residual in hexane extracted flakes, 67, 81
Lipoxygenase, 27, 66, 67, 80
Lysinoalanine, formation in soybean proteins, 86, 87

M

Margarines, 19-22, 73, 74
  blended fats for, 20
  tub, 20, 21, 22, 73
Meat products and analogs, soybean proteins in, 42, 43, 44, 46, 47, 62, 63, 64, 88, 89
Miso, see Oriental soybean foods

N

Natto, see Oriental soybean foods
Nitrogen solubility index (NSI), 30, 31
Nutritional properties of soybean proteins, 51-58, 68, 75, 76, 85, 86, 87
  antinutritional factors, 51-54, 75, 76, 85
  cereal-soybean blends, 85, 86
  essential amino acids, 33, 34, 55, 56
  flours and grits, 55, 56, 85, 86
  oriental foods, 58
  protein concentrates, 56, 57
  protein isolates, 57, 58
  spun protein fiber foods, 58, 86
  textured soy flours, 56, 86

O

Oil extraction, 11, 12
Oil, palm, 70
  deodorization, 76
Oil, soybean
  alkali refining, 14
  bleaching, 14, 78
  consumption, 70, 74
  degumming, 12
  deodorization, 15-17, 76
    equipment, 16, 76
    temperature, 17, 76
    time, 17, 76
  fatty acid composition, 20
  flavor stability, 22-24, 77-79
    iron, 22, 78, 79
    linolenate, 23
    metal inactivating agents, 17, 18, 22, 23
    oxidation products, 23, 79
  from damaged beans, 78, 79
  hydrogenation, 14, 15
    nonselective, 15
    selective, 15
  products
    cooking and salad oils, 18, 19, 20, 77, 78
    copper reduced, 23, 24, 77, 78
    hydrogenated winterized, 18-20
    labels, 74
    nickel reduced, 18-24, 77, 78
    shortenings and margarines, 19-22, 73, 74
    stabilizers, 17, 22, 23, 77
Oilseeds, other, 70
  palm, 70
  sunflower, 70
Organizations, soybean, 72
  American Soybean Association, 72

National Soybean Processors Association, 72
  other, 72
Organoleptic properties of protein products, 65-67, 87, 88
Oriental soybean foods, 58, 59, 60
  kinako, 59
  miso, 59
  natto, 58, 59
  shoyu (soy sauce), 59, 60
  tempeh, 58, 60
  tofu, 58, 59

P

Polyunsaturates and P/S ratio, 20, 74, 78
  Authors' introduction
Processing of soybeans
  desolventizing, 12, 35-38
  extraction, 11-12
  oil and meal, 7-12
  preparation of beans, 11
  soy flours and grits, 7, 8, 35-38, 80, 81
  soy protein concentrates, 7, 8, 39, 81
  soy protein isolates, 7, 8, 40, 41, 82
Protein
  alkali treatment, 29, 86, 87
  amino acid composition, 33, 34, 55, 56
  association-dissociation reactions, 28, 29
  bodies, 3, 5
  chemical properties, 25-34
  denaturation, 30-33
  determination in foods, 90
  dispersibility (solubility), 25, 26, 29, 30, 31
    analytical methods (NSI and PDI), 30, 31
  disulfide polymers, 28, 29
  flavor properties, 65-67, 87, 88
  gel filtration, 27, 28
  gelation, 31, 44, 46
  isoelectric point, 25, 26
  isolation, 40, 41
  lysinoalanine formation on alkali treatment, 86, 87
  manufacturers, 79, 80
  molecular size, 26, 27
  production estimates, 41, 42, 79
  sedimentation in ultracentrifuge, 26-29, 33
  selling prices, 41, 42, 79
  structure, 28, 29
Protein, concentrates, see Soy protein concentrates
Protein, food uses, see Food uses of soybean proteins
Protein, functional properties, see also Functional properties of soybean proteins, 43-51, 67, 68, 83-85
Protein, isolates, see Soy protein isolates
Protein, nutritional properties, see Nutritional properties of soybean proteins
Protein solubility, see Protein, dispersibility

R

Raffinose, 3, 75

S

Salad oils, 18, 19, 20, 77, 78

Saponins, 53, 54
Schneckens desolventizing system, 35, 36
Seed
  color, 3
  composition, 3-5
  grading standards, 7
  origin, 69-70
  parts, 3-5
  production, 3-7, 70, 72-76
    by states, 6, 73
    by year, 6, 73
    early history, 3
    future, 70
    yield barrier, 75
  storage, 8-11. 74, 75
  structure, 2-5
  supply and disposition, 7, 73
  symbol, 69
  varieties, 6-7, 75-76
    breeding, 6, 7, 75
Shortenings, 19-22
  all-purpose, 21
  fatty acid composition, 20
  high stability, 21
  liquid, 20
  solid fat indices, 19, 21
Shoyu, see Oriental soybean foods
Snack foods, soybean proteins in, 61, 65, 89, 90
Soybean oil, see Oil, soybean
Soybean proteins, see Protein
Soy flours and grits
  amino acid composition, 33, 34, 55, 56
  composition, 34, 35
  denaturation of proteins, 30, 31, 35-39
  extruded products, 37, 39, 42-44, 46-48, 56, 61, 63, 79, 80, 82, 84, 88, 89
  flavor, 65-67, 87, 88
  food uses
    baked foods, 42, 43, 44, 45, 46, 51, 60, 61, 62, 88, 89
    cereal products, 42, 61, 64, 85, 86
    dietary foods, 61, 65
    infant foods, 42, 61, 64, 85, 86
    meat products and analogs, 42, 43, 44, 46, 47, 62, 63, 64, 88, 89
  hexane desolventization, 35-38
  manufacturers, 79, 80
  moist heat treatment, 30, 31, 35-39, 51-56, 80, 82, 85, 87
  nutritional properties, 55, 56, 85, 86
  production estimates, 41, 42, 79
  selling prices, 41, 42, 79
Soy flours and grits, functional properties, see Functional properties of soybean proteins
Soy protein concentrates
  amino acid composition, 33, 34, 55, 56
  composition, 39, 40
  denaturation of proteins, 39, 81
  flavor, 65, 87
  food uses, 43, 44, 45, 46, 61, 62, 63, 64, 89
  manufacturers, 79, 80
  nutritional properties, 56, 57
  processing, 39, 81, 82
  production estimates, 41, 42, 79
  selling prices, 41, 42, 79
Soy protein concentrates, functional properties, see Functional properties of soybean proteins
Soy protein isolates
  amino acid composition, 33, 34, 55, 56
  composition, 41
  denaturation, 29-33, 57, 58, 82
  flavor, 65, 66, 67, 87
  food uses
    beverages, 60, 61, 64
    infant foods, 61, 64
    instant breakfast items, 89
    meat products and analogs, 48, 49, 61, 62, 63, 64, 88, 89
    manufacturers, 79, 80
    nutritional properties, 57, 58, 68, 86, 87
    processing, 40, 41, 82
    production estimates, 41, 42, 79
    selling prices, 41, 42, 79
    solubility, 25, 26, 29, 40, 41, 67, 68, 83
    spun fibers, 48, 49, 50, 63, 64, 82, 84
Soy protein isolates, functional properties, see Functional properties of soybean proteins
Spherosomes, 3, 5
Spun protein fibers, 48, 49, 50, 63, 64, 82, 84
Stachyose, 3, 75
Storage of soybeans, 8-11, 74, 75
  for export, 74, 75

T

Tempeh, see Oriental soybean foods
Textured protein products, 46, 47, 48, 49, 50, 71, 82
Tofu, see Oriental soybean foods
Trypsin inhibitors, 51, 52, 53, 85

V

Vapor desolventizer-deodorizer, 35, 38
Varieties of soybeans, see Seed, varieties

W

Waste disposal (whey), 68

Y

Yield barrier of soybeans, 75